Understanding Copper Alloys

UNDERSTANDING COPPER ALLOYS

The Manufacture and Use of
Copper and Copper Alloy
Sheet and Strip

edited by
J. HOWARD MENDENHALL

Olin Brass
East Alton, Illinois

A Wiley-Interscience Publication

JOHN WILEY & SONS

New York • Chichester • Brisbane • Toronto

Library of Congress Cataloging in Publication Data:

Olin Brass (Firm)
    Understanding copper alloys.

    "A Wiley-Interscience publication."
    Includes bibliographical references and index.
    1. Copper alloys. 2. Copperwork. I. Mendenhall,
J. Howard. II. Title.

TS565.044   1980   673'.3   79-24502
ISBN 0-471-04811-9

Printed in the United States of America

10 9 8 7 6 5 4 3 2 1

To John J. Cronin, Jr. whose untimely death precluded his seeing his contribution to this book in final print.

PREFACE

This book was prepared by the Marketing Engineering Department of Olin Brass. It is intended to provide a basic guide and reference source for those who are designing, engineering, or planning the fabrication of some new part from copper or its alloys.

In order to melt and process copper and copper alloy sheet and strip that is tailor-made to fit the user's fabricating methods and provides the properties required in the end product, it is necessary to know the shape of the part to be produced, the methods to be used to fabricate the part, and the function the part is intended to perform. Alloy, thickness, temper, and surface properties are major determinants of how well the metal will perform, and selection is generally based on an understanding of them.

The information contained in this book is meant to furnish background and guidance toward a cost-effective selection process. It is drawn from the many years of practical experience of Olin's application engineers and backed by many standard test data from Olin's plant laboratories and special test data developed in its Metals Research Laboratory.

Preparation of the text was under the direction of P. F. Inman, Manager of Marketing Engineering, with assistance from J. H. Mendenhall. Many Olin Brass personnel contributed to its preparation. Contributors included M. W. Rupp, W. T. Ward, R. J. Slusar, J. A. Grivalsky, B. T. Cox, J. R. Scott, J. H. Hall, K. R. Schweigert, S. P. Zarlingo, J. H. Mendenhall, and the late J. J. Cronin. Special thanks are due to F. Kenneth Bloom, who provided editorial assistance.

P. F. INMAN

Olin Brass
East Alton, Ill.
January 1980

CONTENTS

1 INTRODUCTION   1

Brief description of the purpose and contents of the book

2 MELTING AND MILL PROCESSING OPERATIONS   3

Raw Materials / Casting / Hot Rolling / Milling or Scalping / Cold Rolling / Annealing / Cleaning / Slitting, Cutting, Leveling

3 CLASSIFICATION SYSTEMS FOR ALLOYS AND TEMPERS   43

The UNS Numbering System for Designating Copper and Copper Alloys / Basic Composition of Alloy Families and their Properties and Uses / Temper, Its Meaning and Measurement / Designation of Tempers

4 METALLURGY OF COPPER   51
AND COPPER ALLOYS

Atomic and Crystal Structure / Plastic Deformation and Slip / Effects of Alloying on Structure
Phase Diagrams / Precipitation / Effect of Impurities

5 MECHANICAL PROPERTIES   65
AND GRAIN SIZE

Tension Testing, Stress-Strain Curves, Elastic
Properties, and Yield Strength / Young's Modulus
and Secant Modulus / Elongation / Grain Size,
Definition and Measurement / Hardness Tests,
Methods and Limitations / Tensile Strength and
Temper Designations

6 PHYSICAL PROPERTIES—ELECTRICAL   82
CONDUCTIVITY, THERMAL
CONDUCTIVITY, ELEVATED
TEMPERATURE EFFECTS,
AND DENSITY

Electrical Conductivity, Mechanism / Effect of
Alloy Additions and Impurities / Effect of Temperature / Thermal Conductivity / Thermal Expansion / Density

7 OTHER ENGINEERING PROPERTIES—   94
FATIGUE, CREEP, AND RELAXATION

Effect of Alternating Stress, Fatigue / S-N Diagrams / Elevated Temperature Creep and Stress
Rupture Strength / Stress Relaxation

8 CORROSION RESISTANCE   107

Formation of Oxide Films, Method of Removal
by Chemical Cleaning / Electrochemical Theory
of Corrosion / Atmospheric Corrosion / Aqueous
Corrosion / Galvanic Corrosion / Biofouling /
Erosion-Corrosion / Pitting / Stress Corrosion /
Corrosion Fatigue / Dezincification / Intergranular Corrosion

9 ELECTRICAL CONTACT   132
RESISTANCE

Measurement of Contact Resistance / Test Results
for Various Alloys / Use of Coatings / Metal
Dipping / Electroplating / Electroless Chemical
Plating

10 FABRICATION BY FORMING,   143
DRAWING, AND RELATED OPERATIONS

Principals of Deep Drawing / Alloy Selection /
Effect of Temper / Lubrication / Methods of
Correcting Defects / Other Forming Operations:
Coining, Stretch Forming, Blanking, Spinning,
Hole Flanging or Drifting / Theory of Bending

11 JOINING: WELDING, BRAZING,   190
AND SOLDERING

General Review of Various Joining Techniques /
Welding: Selection of Best Process, Operating
Conditions / Brazing / Soldering: Effect of Alloy
Composition on Solderability Ratings, Surface
Preparation, Fluxes, Methods

12 SURFACE TREATMENTS: COLOR 211
AND COLORING, POLISHING,
BUFFING, PATTERNED METALS

Natural Color of Copper / Effect of Alloying /
Chemical Coloring Baths / Protective Coatings /
Polishing and Buffing, Preliminary Operations /
Classification of Polishing Wheels / Hard and
Color Buffing / Embossed Finishes / Styles and
Specifications

13 SPRING DESIGN AND 228
MATERIAL SELECTION

Design Criteria / Critical Parameters / Mode of
Stressing / Modulus / Fatigue / Relaxation and
Creep / Formulas for Various Types of Springs /
Examples

14 SPECIFYING 261
PRODUCT QUALITY

How Quality Is Described / Specifications /
Dimensional Tolerances for Thickness, Width,
Length, Camber, Flatness / Other Quality Re-
quirements / Defects and Troubleshooting

15 ECONOMICS OF 285
STRIP SELECTION

Cost-Effective Selection of Gauge and Width /
Effects of Density, Coil Length, Anisotropy
(Minimizing Web Scrap) / Formula for Choosing
Most Economic Alloy

APPENDIX—WROUGHT COPPER
AND COPPER ALLOYS LIST

303

INDEX

317

Understanding Copper Alloys

1

INTRODUCTION

Copper and copper alloys in the form of sheet and strip are used in applications that are for the most part functional parts. A basic characteristic of the metal and its alloys is nobility, an inherent resistance to attack or deterioration in natural atmospheres. This characteristic is complemented by strength and formability, which enable these metals to be formed into useful parts and to withstand the levels of stress encountered in many common situations. Strength comparable to that of the common structural steels is found in many copper alloys.

Copper is far and away the common industrial metal with the highest electrical and thermal conductivity. This is also an inherent property of copper which when coupled with its inherent strength and corrosion resistance makes it and its alloys unique in their usefulness as conductors of electricity and heat.

The chemical, physical, and mechanical properties of these metal products are the basis for their usefulness. In the following chapters some of the more important of these properties are described. How properties are controlled in the manufacturing processes, and how the properties are applied to parts fabrication and application, are discussed. The information is intended to provide materials engineers and parts designers with a better understanding of these products and their useful properties.

In Chapter 2 some production methods used in the manufacture of copper and copper alloy sheet and strip are described to provide understanding of how useful properties are developed and controlled by process design. Some limitations and their reasons for being are also revealed in the discussion of process methods. In Chapter 3 the coppers and copper alloys are classified and the classification system is explained. Temper, which is second in importance to composition, is also ex-

1

plained. In Chapter 4 some principles of the physical metallurgy of copper and copper alloys are given to provide understanding of why they have their unique and valuable properties.

Chapters 4, 5, 6, 7, 8, and 9 describe the properties of the materials. In Chapter 5 the mechanical properties are shown to be functions of processing to meet temper specifications. The important mechanical properties and the test methods by which they are determined are described. Chapter 6 describes physical properties and the ways in which they sometimes respond to changing conditions. Chapter 7 acquaints the reader with some of the less common properties whose influence may be determinants of metal choice when service conditions are more demanding. Chapter 8 discusses some corrosion resistance considerations for the application of these metals. Chapter 9 is closely related, describing the influence of surface phenomena on the electrical contact resistance of some copper alloys.

Chapters 10, 11, 12, and 13 are concerned with applications. In Chapter 10 some fabricating techniques frequently used in parts manufacture are described, and alloy properties are related to deep drawing and forming. Chapters 11 and 12 show how the properties of copper and copper alloys make them particularly desirable for finishing and joining to provide both decorative and utilitarian products. Chapter 13 calls attention to the special considerations that are required to take advantage of the usefulness of copper alloys in spring applications. Emphasis is on the design criteria for contact springs and connectors in electrical and electronic applications.

Chapters 14 and 15 cover quality and economic considerations of strip selection. In Chapter 14 a discussion of the dimensional characteristics and their tolerances and other general quality considerations is aimed at assisting the specifications writer and the metal buyer to relate their metal needs to the industry standards, and at a mutual approach to problem solving by user and supplier. Chapter 15 calls attention to some of the special considerations that will assist the user to make the most economical selection of the best material to do the required job.

No effort is made to fully cover any of these subjects in this book, but they are highlighted to assist the metal user. If users and potential users of copper and copper alloy strip and sheet gain information from this effort that helps them to utilize the described products in a technically sound and cost-effective manner, its purpose will be accomplished.

2

MELTING AND MILL PROCESSING OPERATIONS

The manufacture of sheet and strip in the modern brass mill begins with one of two basic casting operations. In the casting plant the metal is melted and either cast in the form of slabs which are subsequently heated and hot-rolled to coils of heavy gauge strip, or directly cast in strip form and coiled. The coils, in either case, will have their surfaces milled to remove any defects from casting or hot rolling. The next set of operations through which they are brought will provide the desired final gauge and temper by a series of cold-rolling, annealing, and cleaning operations. Finally, they may be slit into narrower widths, leveled, edge rolled or otherwise treated, and packaged for shipment.

RAW MATERIALS, MELTING, AND CASTING

Raw materials from which the melt is prepared consist primarily of virgin copper, either electrolytic or fire-refined, selected clean scrap (Figure 2-1)[1] of known origin, carefully checked for composition, and special alloy elements such as virgin zinc, lead, tin, or nickel (Figure 2-2).[1] Scrap is baled to make it dense, so it sinks below the surface of the molten metal in the furnace and melts rapidly. This is the first of many operations in which the quality of the finished product can be affected.

[1]The description and illustrations of processing equipment in this chapter are drawn from the facilities of Olin Brass at East Alton, Illinois, and New Haven, Connecticut, and Somers Thin Strip at Waterbury, Connecticut. They are typical of modern brass strip mills.

FIGURE 2–1
Baling machine compresses high-purity brass scrap. It will become part of a carefully formulated furnace charge.

In this case the impurities that can alter the processing characteristics of the copper alloys can be avoided or controlled to established tolerance levels. Careful screening and control of virgin metals and vigilance in the use of scrap are keys to successful quality control in this part of the operation.

After the charge has been formulated, the raw materials are assembled in charge buckets and carefully weighed. These materials are discharged into hoppers which feed electric-induction melting furnaces (Figure 2–3). As melting of the charge proceeds, samples are taken from the furnace and sent to the spectrographic laboratory for analysis. The composition is calculated by computer and returned to the printer on the melt shop floor within minutes. If necessary, the melter can then make additions to bring the melt exactly within the specified composition range. With this type of melting, the analysis of the raw materials is carefully controlled and few or no impurities are introduced during melting. The metal is protected from atmospheric oxidation by a cover of carbon or bone ash. When the composition and temperature have been determined to meet the requirements of the alloy being melted, the molten metal is transferred to a holding furnace.

For many years in brass mill casting practice the molten metal was poured from the melting furnace into a pouring box which distributed it into long, rectangular book molds (split molds, hinged like a book). The slab cast in these molds was some 2 to 4 in. thick, 2 to 2.5 ft. wide, and about 8 feet long. The book mold method had some important disadvantages. The maximum weight of a casting, and therefore the length of finished coil, were limited. Molten metal dropping 8 ft. from a pouring box to the base of the mold was subject to oxide scumming and entrapment. Metal splash caused the bottom end of the bar to be spongy. The casting varied from bottom to top in temperature and solidification rate with potential problems from shrink cavities, gas entrapment, surface laps, and mold coating defects.

FIGURE 2–2
Virgin zinc, an alloying element, being added to a "charge bucket" to insure that the alloy's composition limits will be met.

FIGURE 2–3
The contents of a charge bucket drop into a vibratory feeder and then are fed into a melting furnace.

During the 1960–1975 period, two *semicontinuous casting* processes began to supplant the book molds. In each of these new methods, molten metal flows into a short, rectangular, water-cooled mold, which initially is closed at one end by a plug on a movable ram or a starter bar (Figure 2–4a). The metal freezes to the plug and forms a shell against the mold surface. The ram is then steadily withdrawn, pulling the shell with it. As the shell exits from the bottom of the mold, cold water is sprayed on it, cooling it rapidly and causing the contained molten metal to freeze. In this manner a continuously cast slab of the desired length is produced.

The direct chill (DC) casting process described above is done in vertical molds and is used to produce slabs of large cross section which are subsequently reheated, hot-rolled into heavy gauge strip, and coiled. A second continuous casting process uses a horizontal mold and casts a thin, rectangular section in much longer lengths which are coiled directly without hot rolling (Figure 2–4*b*).

The DC, vertical semicontinuous cast method is designed to produce large slabs from which heavy weight finished coils can be made. Such large coils are the most economical to handle through the subsequent rolling and annealing processes at the brass mill and later by the user who is fabricating finished parts.

Direct chill slabs are hot-rolled to produce coils, but not all alloys can be rolled by this method. Tough pitch copper castings contain a very small amount of copper oxide and benefit from hot working through having these brittle grain-boundary oxide networks dispersed. On the other hand, some alloys contain elements which produce phases, or structures, which are difficult or even impossible to hot-roll. Such alloys must be cold-rolled, and the amount of reduction in thickness that can

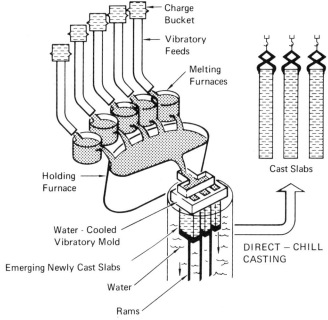

FIGURE 2–4*a*
Schematic sketch of vertical direct chill (DC) semicontinuous casting of slabs.

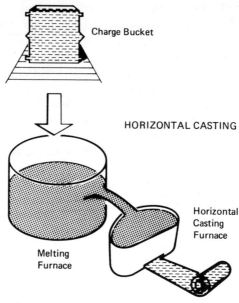

FIGURE 2–4*b*
Schematic sketch of horizontal continuous casting and coiling of strip.

be achieved, before annealing becomes necessary, is small when compared to hot-rolling reductions.

The problem with alloys that are hard to hot-work is overcome with the horizontal continuous-casting method. It offers a means of producing relatively thin castings in long lengths which can be coiled in the cast state and later reduced by cold rolling. Tedious, costly cold breakdown rolling and the attendant annealing are avoided.

In the DC casting process, more than one slab can be cast from each pour. A typical DC casting station consists of multiple melting furnaces and a large holding furnace. When the metal in each of the melting furnaces has been melted, the composition has been established, and the proper temperature attained, the molten metal is transferred to the holding furnace. Casting then proceeds from the holding furnace. Typically three 15,000 lb. slabs of the same uniform composition are cast at one time. Smaller versions of this same process exist in the industry as well.

During pouring, samples are again taken and analyzed in the spectrographic laboratory (Figure 2–5). This analysis will be the official final analysis of the slabs and subsequent coils produced from the casting. The furnace is elevated and tilted forward as casting is begun. The molten metal flows through a narrow channel and is fed into a pouring box and then through a spout into the mold. The mold is short, 8 to 20 inches, so the metal can be introduced smoothly. The mold is oscillated slightly to keep the freezing metal from adhering to it and to assist in removal of the slab. After the mold is full and the ram begins withdrawal of the slab, the end of the pouring spout is kept below the molten metal level to eliminate splashing and to prevent oxidation. The pouring box contains valves which regulate the rate of flow and coordinate it with the withdrawal rate. These two factors, as well as the rate of flow of cooling water, are controlled carefully. Each alloy has a different solidification rate and requires slightly different casting conditions.

The relatively small molten pool and the rapid and uniform cooling promote a homogeneous, fine-grained cast structure, which is very beneficial to hot workability. In addition, the continuous upward freezing of the molten pool allows gases to rise and be freely expelled. Any nonmetallic material also floats to the surface of the shallow pool and

FIGURE 2–5

Samples from the melting and holding furnaces are sent by pneumatic tube to a spectographic laboratory, where direct-reading spectographic and X-ray flourescence equipment analyze the specimens in less than 2 minutes, so the composition of the metal being cast can be controlled and recorded.

remains there to collect at the extreme top of the cast slab. This gate end of the casting is then sawed off before hot rolling, leaving a product of unusual cleanness and high quality (Figure 2–6).

The horizontal continuous-casting process also provides a product of excellent quality. Typically, one low frequency electric-induction furnace is used as a melter. As a charge of selected scrap and refined metal additions are melted and brought to the pouring temperature, samples for chemical analysis are taken. When the proper analysis is established and the pouring temperatue attained, part of the metal is poured into a second, smaller electric-induction holding furnace. This furnace is constantly monitored to maintain the metal at the desired casting temperature. The casting mold is attached to the lower front of this furnace. It is a graphite mold contained in a copper, water-cooled jacket. A silicon carbide plate in the front of the furnace contains a slot which opens into the mold. At the beginning of a cast, a starter bar is inserted into the mold and the metal freezes to it. The mold is only a few inches long.

Two stands of withdrawal rolls slowly withdraw the starter bar as the metal freezes in the mold cavity. The cast bar, frozen to the starter bar, is continuously withdrawn as the metal freezes in the mold. Although it is a simple process, its practice requires that tolerances on mold dimensions be held to a few ten-thousandths of an inch, and exceptional melt cleanness be maintained. Any dross or other foreign material that enters the graphite mold will quickly destroy it. Mold sizes range from 10 to over 24 in. in width and from about ⅜ to ⅝ in. in thickness. A saw in the withdrawal line cuts the bars off at the desired length, and they are coiled in preparation for subsequent processing. A sample for chemical analysis is cut from each bar end, so the composition at each end of each coiled bar is determined. This process lends itself to in-line coil milling and to maximum coil lengths, dependent only on handling equipment capacity and practical processing of the material itself.

The rapid chilling of the small amount of metal in the horizontal mold produces a fine, equiaxed cast grain structure. The metal drawn from the furnace as it solidifies always has a pool of molten liquid above it where gases and nonmetallic impurities tend to collect. The cast bar is free of porosity and of defects caused by solid inclusions.

The good quality of the horizontal casting shows up in the finished strip in terms of excellent formability. Phosphor bronzes cast this way develop the high strength for which they are specified, coupled with the good formability needed in most of their applications. Leaded bearing-bronze Alloy C54400, also cast by this process, offers improved quality for bushings, bearings, and thrust washers which must carry heavy loads under dynamic stresses without failure. Some smaller mills depend

FIGURE 2–6
A direct-chill-cast slab is removed from the mold. The length of these slabs is 25 ft, and cross sections may be as large as 32 in. x 6 in.

almost entirely on horizontal casting, regardless of alloy, because the process is readily adaptable to the casting of small quantities of several alloys.

Both continuous-casting processes provide a cast product which is fine grained, sound, and free of nonmetallic inclusions.

HOT ROLLING OF DIRECT CHILL CONTINUOUS-CAST SLABS

Sheet and strip are *wrought products* by definition. Their properties differ from those of the cast slab because their grain structure has been refined. The initial rolling of slabs is for this purpose as well as to begin reduction in thickness.

For copper and copper alloys that can be hot-worked, the quickest and most economical method of reduction is hot rolling. When metals are worked cold, as in cold rolling, they work-harden. As a result, the cast structures of most alloys can only be reduced in thickness approximately 20 to 40% by such means before they become too hard and brittle for further cold rolling. While annealing will produce recrystalization and a softened wrought grain structure, to reduce a cast bar 6 in. thick and 32 in. wide to a thickness of 0.400 in. would require many cold-rolling and annealing operations. The annealing furnace for the last anneal would have to approach 500 ft. in length. This is a difficult and expensive way to produce strip or coiled sheet.

In contrast, by heating the 25 ft. long slab to the appropriate temperature, which can be done in a furnace of reasonable size, and hot rolling, the reduction can be accomplished in a single heating operation followed by a series of passes through a hot-rolling mill. This is possible, because hot rolling is done at a temperature that simultaneously causes recrystallization. As the grain structure is being deformed, recrystallization occures (Figure 2–7). The metal does not work-harden as long as its temperature remains above this point, and reduction in thickness by rolling can proceed. The hot-rolling process is scheduled so the metal will be rolled to the desired thickness before it cools below the minimum hot-working temperature.

To ready the slab for hot rolling, the top or gate end is trimmed by sawing, and then it is conveyed into a furnace for heating. Slabs or bars of the same alloy are grouped together in a lot and processed through the furnace and the hot mill. The furnace temperature and the time for each bar to pass through the furnace are adjusted in order to allow the bar to reach the appropriate temperature throughout its thickness, length, and width by the time it passes through to the exit conveyor (Figure 2–8).

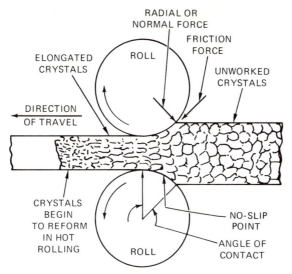

FIGURE 2–7
A sketch illustrating the hot rolling process.

FIGURE 2–8
Delivery of a hot slab to the entry table of the hot mill from the discharge end of the slab heating furnace.

Temperature control is an important factor in hot rolling (Figure 2–9). Hot rolling can be accomplished only within a certain temperature range for each alloy. The bars will be damaged and have to be scrapped if hot rolling is attempted at a temperature which is too high or too low. Further, for all alloys, the grain size of the hot-rolled product is determined by the temperature at the last rolling pass. Subsequent processing to meet specified properties is dependent on this grain size.

In some alloys, as noted in Chapter 4, elements go into solution above certain temperatures and then precipitate out at lower temperatures. By completing hot rolling at a temperature above the precipitation temperature and quenching in a high pressure water spray, solution heat treatment can be accomplished. This affects both the physical and the mechanical properties attained in subsequent processing.

The roll stand used for hot rolling is a very sturdy mill having two rolls (*two-high*) whose direction of rotation can be rapidly reversed so the strip can be passed back and forth between them. The large horizontal rolls

FIGURE 2–9
Slabs for hot breakdown rolling are heated to approximately 850°C (1560°F). A technician checks the furnace instrumentation.

which reduce the thickness are supplemented by a pair of vertical edging rolls. Hot rolling is the only rolling operation on strip in which any appreciable spread in width takes place. The vertical rolls are needed to maintain the proper width by rolling edges.

The rolls are water cooled to avoid overheating, which would cause the surfaces to crack and check. Further, a polishing stone continuously dresses the rolls as they operate. As the thickness is reduced, the bar length increases proportionately. A 50% reduction in thickness doubles the bar length. In order to handle a 25 ft. cast bar 6 in. thick a run-out conveyor on one side of the roll of 350 ft. length is required. This is needed to accommodate the bar on the next to the last rolling pass. On the other side of the roll stand is the high pressure water spray and a coiler. After the final rolling pass the metal is spray cooled and coiled.

The modern hot mill is operated from an air-conditioned pulpit overlooking the rolling stand and the conveyor run-out table (Figure 2–10). With the aid of television cameras, placed at strategic points

FIGURE 2–10
The hot-rolling operation is monitored visually from a control pulpit (a), as well as by readout gauges and closed circuit television cameras (b). The metal is cooled by water jets and then put into coil form by a power coiling device at the end of the "run-out" table (c).

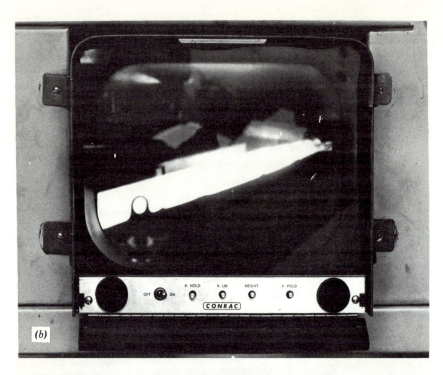

(b)

(c)

16

focusing on the rolls, the furnace and transfer buggy, the alligator shears at each end, and the coiler, the operator can control all these from his vantage point. The shears are used for trimming bar ends as required. Rolling temperatures and the percent of reduction per pass are designed to suit each alloy. A schedule of rolling reduction for each pass through the rolls is established and recorded on a punched card. Operation of the hot mill is sequenced by computerized controls to insure uniform processing through the hot mill.

MILLING OR SCALPING

Along with continuous casting, an equipment development that significantly advanced process methods was the *high speed coil milling machine.* All coppers and copper alloys, produced with the good surface expected of brass mill sheet and strip, have their surfaces removed or *scalped* by a machining operation after the breakdown rolling to remove all surface oxides remaining from casting or hot rolling. This operation is accomplished in a specially designed milling machine having rolls with inset blades which cut or mill away the surface layer of metal. The capability of this machine to handle the product in coiled form means that a much longer bar can be conveniently and economically milled (Figure 2–11).

Following hot rolling the DC cast bars are coil milled, and after careful surface inspection are ready to be applied on orders for processing to final gauge, temper, and width. Horizontally continuous-cast bars arrive at this stage by a somewhat different processing path. The coiled cast bars are annealed to provide a stress-free structure of maximum ductility. They are then cold rolled to work the structure sufficiently so a fully recrystallized wrought grain structure will develop in the subsequent anneal. The bars are then scalped by milling.

Both hot-reduced and cold-reduced milled bars are in the thickness range of 0.300 to 0.400 in.

COLD ROLLING TO FINAL THICKNESS

The sequence of operations for processing metal from milled condition to finish thickness or gauge is designed to meet specified requirements for each particular application.

All thickness reduction is accomplished by cold rolling, and a variety of rolling mills are used. Cold rolling of coppers and copper alloys into sheet and strip of excellent quality requires a combination of skillful workmanship, knowledge, and good rolling mills. To keep cost as low as

FIGURE 2–11
A coil miller. Entry end is at the right. The miller uncoils, flattens, and mills the top, bottom, and edges of hot-rolled strip in one pass, recoiling the metal as it emerges.

possible and competitive, the reduction in thickness to final gauge needs to be accomplished in the fewest operations compatible with quality requirements. The basic problem is to reduce the thickness as much as possible in each rolling operation, while maintaining uniformity of thickness across the width and from end to end of a coil which is 200 to 600 ft. long at the first rolling pass and could end up 25,000 ft. long if rolled to 0.004 in. finish gauge. Coupled with the need to maintain uniformity of thickness through all processing stages is the need to maintain flatness across the width and along the length of the coiled metal. Metal with uniformity of flatness across and along its length is described as having "good *shape*." It is free of "humps," "waves," and "buckles."

A rolling mill is capable of applying a large, but still limited, force upon the surfaces of the metal as it passes between the rolls to reduce its thickness. The applied force is spread across the contact area of the rolls on the metal. The larger the contact area, the smaller the force that is applied per unit of area and the smaller the reduction in thickness per pass through the rolls that can be achieved. Rolls of small diameter will have a small contact area, and greater force per unit of area. Small diameter work rolls are most desirable for providing maximum utilization of roll force in reducing metal thickness, but they lack the stiffness required. The wider the metal to be rolled, the longer the rolls, and the greater the tendency for the rolls to bend or spring. To overcome the tendency *four-high and cluster rolling mills* are used for cold rolling in the brass mill.

Four-high rolling mills (Figure 2–12) contain a pair of work rolls of relatively small diameter (e.g., 12 in.). A second pair of rolls, of large diameter (e.g., 36 in.), is placed above and below the work rolls in the stand to back them up and prevent them from springing. This arrangement allows the advantage of the small contact area of small work rolls and the transmittal of high force through the large back-up rolls while maintaining the rigidity required for gauge control. The minimum size of the work rolls is limited by the forces in rolling, which tend to bow them backward or forward during rolling.

Cluster rolling mills, for example, *Sendzimir* or "Z" mills (Figure 2–13), were designed to counteract both the vertical and horizontal elements of the rolling forces and thus enable the use of minimum diameter work rolls. In cluster mills the work rolls are backed up by a cluster of back-up

FIGURE 2–12a

The tandem mill is two four-high mills operating in series for the rolling of heavy gauge coils as they are brought out of milled-metal storage.

FIGURE 2–12b
Close up of tandem mill showing the two four-high rolling mills in tandem. The large coils of brass are being conveyed back to the entry end for a second pass.

rolls placed with respect to the work rolls so they contain the rolling forces and prevent bending or springing of the work rolls.

By the use of such rolling mills, the thickness from edge to edge across the width of the 24 to 36 in. wide metal coils can be kept uniform through each gauge reduction by rolling. This edge to edge gauge control contributes to the maintenance of good shape. Good shape contributes to the production of flat, straight metal when slitting to the final specified width needed by the consumer. The control of thickness from end to end of a metal coil, which may be several thousand feet long, is a further requirement of the rolling processes which reduce the gauge. Basic mechnical features of the rolling mills, such as strength and rigidity of the roll stand or housing, the roll bearings, and the motors and drive trains, are beyond the scope of this work. The rolling mill must meet design requirements covering such features to perform well in rolling the metal in the gauge range for which it was designed.

FIGURE 2–13

The design characteristics of this Sendzimir cold-rolling mill permit severe reductions of the strip while maintaining uniformity of gauge across the width and throughout the length of a coil. The two smallest rolls in the center of the cluster of back-up rolls are the work rolls.

The control equipment included in the rolling mill is a feature which bears directly on control of the gauge from end to end of a coil of metal during rolling. For thickness control during high speed rolling, continuous measurement of this dimension is a necessity. Rolling mills are equipped with X-ray and beta-ray instruments (Figure 2–14), which continuously gauge the metal and provide a continuous readout of thickness. There are also control devices which actuate the screws in the roll housings and automatically open or close the gap between the work rolls to adjust the thickness being produced as required. These gauges may also adjust back tension and forward tension applied by payoff and recoil arbors to effect changes in the thickness of the rolled metal.

Rolling also exerts considerable influence on the surface quality of the metal. Work rolls are made of hardened steel, much harder than the

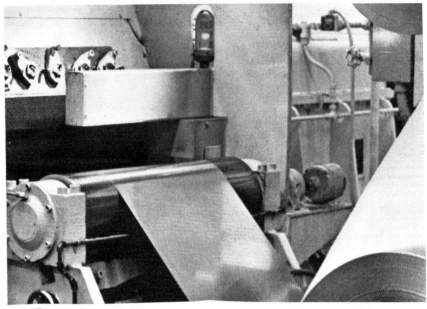

FIGURE 2–14

Beta-ray and X-ray gauges continuously measure thickness during the rolling operation. Shown above is the X-ray gauge employed on a 36 in. wide Sendzimir mill.

copper alloy being rolled. As the rolls squeeze the metal to reduce its thickness, slip between the rolls and the metal surfaces takes place, in both the forward and backward directions. The frictional forces between the roll and metal surfaces, if direct contact were made, would tear the surface of the metal and load the roll surfaces with bits of the metal. To avoid damaging the surfaces in this manner, the metal and roll surfaces are flooded with lubricants to provide a cushion between them. The selection of roll lubricants which will provide the protection needed without staining the metal, will be readily removable from the metal surfaces, and will not interfere with the rolling mill performance is an important engineering function that influences the economic production of high quality copper alloy strip.

The work rolls are meticulously ground to the exact shape for each type of rolling operation and rolling mill design. A pair of rolls 12 in. in diameter with a working surface 30 in. in length may be ground so the diameter in the middle is 0.001 in. larger than at the ends. This difference or crown in the rolls allows for the spring caused by the rolling forces during rolling. The roll surfaces are ground smooth enough so that with the aid of the mill lubricants there will be no dimples or projections on the rolled metal surface. A fine grinding pattern on the rolls helps to hold the lubricant and prevent friction damage to the metal by the rolls. A well-equipped, air-conditioned roll grinding shop, with close temperature control and uniformity so that precise roll dimensions can be attained, is a valuable asset to a brass mill (Figure 2–15).

The more the metal is cold worked, the harder and stronger it becomes. The hardening that occurs when copper and copper alloys are cold rolled allows each of them to be produced with a range of strengths or tempers which are suitable for a variety of applications. Starting with annealed temper, the metal will increase in strength approximately proportionally by the amount of reduction by cold rolling. A series of standard cold-rolled tempers for each copper and copper alloy has been established. A typical plot of reduction versus tensile properties and

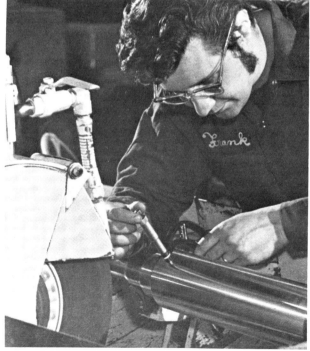

FIGURE 2–15
A skilled technician checks the dimensions of a mill roll in a temperature-controlled roll grinding shop.

hardness is shown in Figure 2–16 for copper Alloy C26000, cartridge brass. The properties of the rolled tempers and their names will be reviewed in later chapters of this book. At this point, it is significant to note that in the processing of strip by rolling to the final thickness both the temper required and the work hardening characteristics of the alloy determine the necessary steps.

For each of the coppers and copper alloys there are limits to the amount of cold reduction that is desirable before annealing the metal to provide a recrystallized soft structure for further cold reduction. Some alloys, such as the phosphor bronzes, the high-zinc-content nickel silvers, and the aluminum-containing high zinc brasses work-harden rapidly. As they are cold rolled, they quickly become too hard for further reduction and must be annealed.

With large amounts of cold reduction prior to annealing, some coppers and copper alloys will develop differences in their strength and ductility when these properties are measured along the direction of rolling, compared to measurements across the direction of rolling. This directionality in mechanical properties arises from the fact that the normal random orientation of the atomic planes from grain to grain is gradually forced into a pattern conforming to the constant working of the metal in one direction. As will be noted in later chapters, this directionality can affect the fabricability and final performance of the

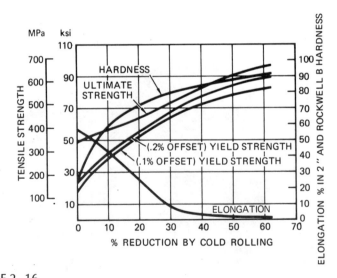

FIGURE 2–16

The effect of cold rolling on the strength, hardness, and ductility of annealed Alloy C26000 when it is cold rolled in varying amounts up to 62% reduction in thickness.

strip or sheet. Its control requires that the amount of reduction between anneals and the temperature of successive anneals be carefully controlled.

The amount of cold reduction that can be made in each pass through the rolls is limited by the inherent strength and power of the rolling mill and the growing resistance of the alloy being rolled to deformation. As the strip thickness becomes less in proportion to the work roll diameter, more force is needed to provide more reduction. The amount of pressure the roll lubricant can stand is yet another limitation on the amount of reduction per pass. If the rolls break through the lubricant film, they can damage the metal surface and also rapidly overheat and damage the rolls themselves. Overheating causes the metal surfaces to become discolored by oxidation and may even produce microscopic hot tears which could be detrimental, depending on the subsequent processing and final application.

Within the foregoing limits, the earliest stages of cold rolling and annealing are designed to achieve the largest practical reduction in thickness. In the two final cold-rolling operations, where the strip is brought to finish gauge, the cold reductions are designed to meet the specified temper requirement. Meeting the tensile strength requirement, which is the basic mechanical property requirement for rolled tempers, is accomplished by cold rolling to the appropriate ready-to-finish gauge, annealing to the desired grain size, and then rolling to finish gauge. The percent reduction between ready-to-finish and finish gauge is chosen to provide the amount of work hardening needed to produce the tensile strength required. Unavoidable small variations in thickness at both ready-to-finish and finsh gauges and in grain size from the ready-to-finish anneal require that the tensile strength requirement be given as a range, rather than a single value.

ANNEALING

Sometimes in the metal working industry any heating and cooling process has been called *annealing*. In the brass mill industry annealing, more specifically, is heating for the purpose of softening and is accompanied by either partial or full recrystallization.

During cold working, grain structure is considerably distorted as slip takes place within the grains. It is this distortion of the structure that is responsible for the hardening and strengthening which occur. One reason for annealing is to soften the metal so it can be further reduced by cold rolling. The metal is heated to a temperature above its recrystallization temperature. At this temperature the atoms in the distorted crystalline structure use the energy stored in the lattice from cold working plus energy absorbed from heating to rearrange themselves into orderly crystalline patterns. Starting from submicroscopic nuclei, new unstrained grains are formed. The metal is again soft and ductile and ready for further working. The size of the new grains depends on the annealing temperature, the time at temperature, the amount of cold work preceding the anneal, and the grain size established by the previous anneal or from hot working. These four factors must be considered when an annealing operation is scheduled.

Anneals are usually designed to produce a chosen grain size, which is measured under the microscope (Figure 2–17). Sometimes, however, the desired grain size is too fine to measure readily. Then the anneal is designed to produce a specified tensile strength, since this property in annealed copper and copper alloys is largely dependent on grain size, with few exceptions. Grain size control is important because, in general,

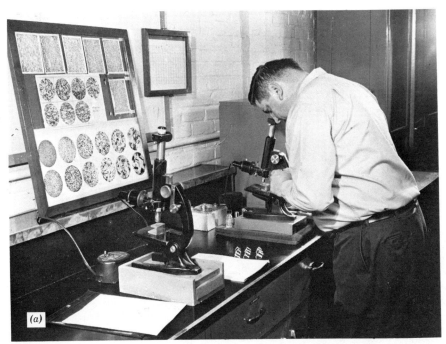

FIGURE 2–17a
Annealed grain size is measured microscopically by a quality control technician.

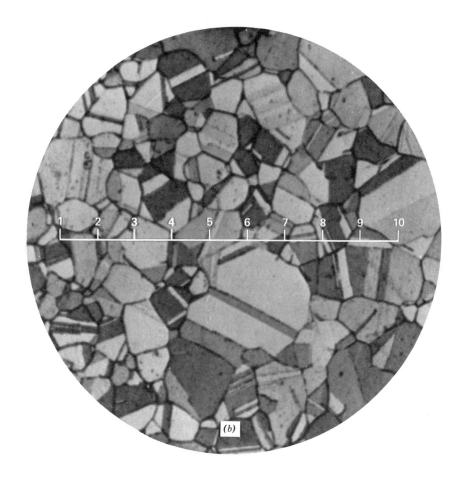

FIGURE 2–17*b*
A typical specimen as viewed through the eyepiece; the scale is for measuring the observed average grain diameter.

the larger the grains the less they interfere with one another when they are deformed. Strain occurs through slip within the grains, and when there are few grains per unit of volume, slip takes place more readily. In addition, less stress or force is required to strain the metal to the braking point. Grain size controls strength and workability. It also contributes to the control of directionality, and it can affect surface roughness. All these factors are considered when selecting the grain size to be established by any of the anneals included in the processing of each coil.

Since grain size is important to process control, it follows that uniformity of grain size is also important. Uniformity can be influenced by the type of annealing furnaces and the method of operation. Each type has certain advantages and disadvantages.

When annealing coiled metal, heat from the furnace must be absorbed through the coil surface and then penetrate to the innermost wraps, mostly by conduction. Temperature will tend to vary in the coil with distance from the heat-absorbing surfaces. *Coil annealing* must be carefully controlled, by slowly applying heat at a rate that will avoid overheating the surface while the temperature of the inner wraps rises and equalizes with that of the outer wraps.

Coil annealing may be carried out in a roller hearth furnace in which the coils are continuously conveyed slowly through the furnace as they are gradually heated to the annealing temperature. This type of furnace usually does not have a prepared atmosphere, but the products of combustion fill the furnace and reduce the metal oxidation rate. More commonly, coil annealing is done in *bell furnaces* of the type in which a controlled atmosphere can be maintained. The annealing unit consists of a base on which the coils are stacked. Under the base is a fan for circulating the hot gases through the load, to provide more uniform and rapid heating. Surrounding the base is a trough, which may be filled with water, oil, or some other material to seal the inner hood when it is placed over the metal load to enclose it for atmosphere control.

In this type of *batch annealing,* bell furnaces capable of annealing up to 100,000 lb. of metal at a time are used (Figure 2–18). After the metal is stacked on the base, temperature control themocouples are placed throughout the load to continuously measure the temperature. The inner hood or retort is placed over the load and sealed. The controlled atmosphere begins to flow through the hood purging the air. The furnace is placed over the hood and heating is begun.

In the well-equipped brass mill large groups of such annealing units may be connected to a central process-control computer. As the furnace and load themocouples measure the temperatures and relay them to the control unit, the heat input is constantly adjusted to maintain temperature uniformity in the load. This controlled temperature rise also allows roll lubricants to vaporize and be carried off before the metal gets so hot that surfaces can be harmed. After the metal has reached the annealing temperature, it is held there for a short period or soaked to provide maximum uniformity. Then the furnace is turned off and removed, and the metal cools in the controlled atmosphere under the inner hood.

FIGURE 2–18
(a) Coils are loaded by crane onto a bell annealing furnace base. (b) Its cover is then lowered into place. (c) The heating furnace is placed over the bell for annealing.

Cooling may be aided by a cooling cover containing a water spray system. The inner hood is not removed until the metal temperature is low enough that no discoloring or oxidation of the metal takes place.

The *controlled atmosphere* is produced in gas cracking units. Combustible gases are burned with sufficient air to oxidize all the gaseous elements. The products of this combustion are then refined, and all gases that would be harmful to the metal surfaces are removed by chemical means. Those remaining pass into the annealing hoods, where they expel the air and protect the metal during annealing. For most coppers and copper alloys an atmosphere which is slightly oxidizing is desirable. For copper Alloy C11000, the atmosphere must be nearly free of hydrogen and the annealing temperature low enough to avoid hydrogen embrittlement. For alloys containing zinc, the small amount of oxygen in the atmosphere combines with the zinc fumes given off and

(b)

(c)

FIGURE 2–18   (continued)

prevents them from attacking the metal parts in the annealing unit. The oxide film which forms on the surface is very thin and readily removed in the subsequent cleaning processes.

One of the advantages of coil annealing in a controlled atmosphere furnace is that the surface of the metal can be readily restored to its natural color by appropriate cleaning following the anneal. The rather rare exception is the case in which an abnormally high annealing temperature is required that causes excessive oxidation or dezincification of a high zinc brass. Special cleaning methods which remove surface metal are then required to correct this condition. The more common situation is that annealing is done in a well-controlled atmosphere followed by normal cleaning practices. This produces a metal surface which is uniform in color and free from detrimental oxides.

A disadvantage of coil annealing is that large coils of some alloys in thinner gauges can be easily damaged. When the coiled metal is heated, it expands and the coil wraps get tighter and tighter. One wrap can become welded to the next because of the high temperature and pressure encountered, usually making the coil unsuitable for further processing. Another disadvantage of coil annealing is that it is time consuming. A large bell furnace full of metal may require from 24 to 40 hours to complete an annealing cycle, and added to this is the time needed for cleaning, done as a separate operation.

In the late 1940s *continuous strand* or *strip annealing* lines began to be used in brass mills. From these early beginnings, the high speed vertical strip annealers were developed in the 1960s. Annealing lines of this type are now in use for annealing copper and copper alloy strip in thicknesses from under 0.010 to over 0.125 in. When several such lines are available, a variety of thickness ranges can be rapidly annealed, providing great flexibility in production scheduling and enabling fast delivery of finished strip.

The continuous strip anneal lines (Figure 2–19) include payoff reels, a stitcher for joining the front end of a coil to the trailing end of the one preceding it, a degreaser for removing roll lubricants, looping towers for metal storage, a seven-story-high vertical furnace which includes heating zone, a controlled atmosphere cooling zone, and a water quenc tank. This is then followed by acid cleaning tanks, a water rinse, a drying oven, and a reel for recoiling the metal. The fact that the metal is uncoiled before passing through the furnace removes any annealing limitations on coil length. Degreasing units remove roll lubricants from the metal surfaces before the metal enters the furnace, so a clean, uniform surface is presented for annealing. The metal passes over a large roller outside the furnace at the top and does not touch anything inside while it is being heated. It then passes under another large roller

FIGURE 2–19
A high speed, continuous strip annealer.

at the bottom in the cooling water tank. This arrangement avoids any possibility of surface damage to the hot metal, which was common in the earlier horizontal strip anneal furnaces. Although the furnace temperature is high, the metal is exposed to it for only a few seconds.

The furnace atmosphere may consist of hot burned gases which are blown against the strip surfaces to heat the metal. The metal is raised to the annealing temperature rapidly and uniformly as it passes through the heating zone of the furnace, and is then cooled rapidly by cold burned gases as it passes through the cooling zone, still protected from excessive oxidation.

Following a water quench, which completes the cooling cycle, the metal passes through the cleaning tanks. A normal cleaning solution is dilute sulfuric acid, which dissolves most of the oxide film left on the metal by annealing. As noted earlier, the atmosphere in the furnace must be slightly oxidizing to prevent zinc fumes from attacking the furnace steel framework. For most coppers and copper alloys this small amount of surface oxidation is not detrimental after normal cleaning, and they are regularly strip annealed throughout processing, including finish gauge. They have a faintly different color than does bell annealed and cleaned metal, but the difference is so slight that it is insignificant in most applications. In fact, brasses containing 15% or more of zinc have surfaces which many users feel are better suited for later fabricating if the strip has been continuously annealed. The metal surface holds lubricants well and has a low coefficient of friction against tool steels, making it desirable for press forming and deep drawing. It is likely that some zinc oxide remains on the surface and acts as a natural lubricant.

After acid cleaning, rinsing, and drying, the surface is usally coated with a detergent solution or a light sulfur-free oil to protect it during handling in transit.

The basic purpose of annealing is recrystallization and softening to prepare the metal for further cold working in the mill or by the consumer. The consistent performance of the metal in subsequent cold working is dependent on grain size uniformity. Since every foot of a coil is exposed to the same temperature as it passes through the strip annealing furnaces, grain size from end to end can be expected to be uniform. Furnace instrumentation continuously records the furnace temperature and controls the heat input. Strip speed through the furnace is similarly monitored. The combination of furnace temperature and speed determines the temperature attained in the metal and, therefore, the grain size. Samples commonly are cut from each end of each coil after strip annealing or coil annealing, and the grain size or mechanical properties are determined as a further control on the quality uniformity of the product.

There are other reasons for heating metal than annealing. *Stress relief* heat treatments are sometimes required. During cold rolling to produce the harder tempers such as Extra Hard, Spring, and Extra Spring, considerable force is required to decrease the thickness and elongate the metal as required. Although the internal residual stresses left in the strip from edge to edge and along the length from this severe working are relatively uniform, small variations sometimes exist which can cause a difference in spring-back during subsequent forming operations. To reduce such residual-stress variations the metal is heated to a temperature below the recrystallization temperature, usually between 200 and

350°C (390 and 660°F), and held there for 0.5 to 1 hour. Such treatment results in a product with uniform spring-back.

Heating for stress relief also can change other properties. In phosphor bronzes tensile elongation is increased and strength slightly decreased. These changes are an advantage in the case of difficult-to-form parts requiring maximum strength. In the high zinc alloys, stress relief heat treatment increases strength and decreases tensile elongation. In this case, the formability may be decreased.

CLEANING

As noted, following each anneal or heat treatment the metal is cleaned. After cleaning in the appropriate solution the metal is thoroughly washed in water, including brushing when needed with wire or synthetic brushes. The rinse water usually contains a tarnish inhibitor, such as tolutriazole or benzotiazole, to protect the metal. For product at finish gauge the rinse tank has a detergent solution added which further protects the metal when dried and also lubricates it slightly to reduce the danger of friction scratches during coiling and uncoiling. Squeegee rolls are used to remove the bulk of the rinse water, and drying ovens in the cleaning lines complete the job.

If desired for subsequent working, annealed strip can also be coated with a film of light nontarnishing oil for protection and lubrication. Metal which is finished in a rolled temper will normally contain a light film of rolling lubricant on the surfaces to protect and lubricate it during coiling and uncoiling and in transit.

SLITTING, CUTTING, AND LEVELING

Following the final rolling, if the metal is to be supplied in a rolled temper, or after the final annealing and cleaning, if it is to be supplied annealed, it is *slit* to its final width (Figures 2–20 and 2–21). During prior processing it may have even been slit to less than full width to meet special requirements. Whatever the preliminary sequence, it is usually slit to final width just before packing for shipment.

Slitting has always been one of the most demanding jobs performed in the strip mill. Slitting cuts the processed coil as precisely as possible into the final width required by the user. Slitting is accomplished by opposing rotary discs mounted on rotating arbors. These knife sets mesh together as the metal passes between them and shear it into a multiplicity of

FIGURE 2–20
Full-width strip is being slit into three finished strips on a medium gauge slitter.

widths. Frequently each width is only a small fraction of the total width of the process coil. The good product yielded from the final slitting operation is the goal of all the processing which has preceded it.

Slitter knife sets are assembled on arbors by skilled craftsmen. The sets are assemblies of disc knives, cylindrical metal and rubber fillers, and shims. The clearance between the opposing knife edges must be exactly correct for the thickness, alloy, and temper of the metal to be slit. The distance between knife edges on each arbor must be set accurately to cut the specified width within the tolerance allowed. Knife edges must be sharp and continuously lubricated. Dull knives or incorrect clearance between knives for the particular material being slit causes distorted or burred edges.

Camber, that is, departure from edgewise straightness, has often been attributed incorrectly to poor slitting practice. It is true that strips can be

FIGURE 2–21
Strips leave the slitter and are rewound and removed individually from the arbor by mechanical "pushers", and delivered to a conveyor, and then automatically banded.

pulled crooked when slitting a large number of them from a wide bar, since the slit strips are sometimes fanned out for subsequent coiling using divider plates. This difficulty is diminished on slitters equipped with over-arm separators because strips need not be fanned out as much. This kind of problem can be anticipated and, if necessary, the bar split at an intermediate stage in processing prior to the final slitting, so fewer cuts are made in this last operation. Tolerances for camber are discussed in Chapter 14.

Actually, rather than slitting practice, it is the maintenance of good shape during each of the rolling operations that is most important in the control of camber. If good control of thickness across the width is maintained at each rolling operation, the edges and centers of the bar will have elongated uniformly, and when narrow strips are slit they will remain satisfactorily straight.

The shape of the slit edge of strip depends to a great extent on the properties of the metal being slit. The metal may be thick, soft, and ductile, at one extreme of shearing characteristics, to thin, hard and brittle, at the other extreme. Between fall all the variations which are characteristic of the gauge, alloy, and temper required for the final application. There is no way that a certain amount of edge distortion can be avoided when slitting thick, soft metals, for example, soft copper, 0.062 in. thick. Even with the best slitter setup, the cross section of a narrow strip will tend to have a "loaf" shape (Figure 2–22). By contrast, thin, hard phosphor bronze or nickel silver in narrow widths will have a cross section of rectangular shape with square cut edges. Leaded brasses shear cleanly because the lead, present as microscopic globules, lowers

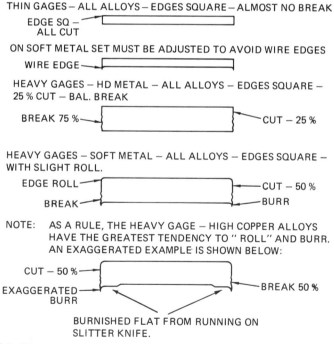

FIGURE 2–22
The different edge contours that can result from slitting, depending on thickness, temper, and alloy.

the ductility and shear strength. It is for this purpose—ease of cutting and machinability—that lead is added to copper alloys.

As the metal comes from the slitter, both edges of each strip, if distorted, will be distorted in the same direction. The immediately adjacent strips will have edges distorted in the opposite direction. There are some applications for which it is desirable that any edge distortion be in the same direction relative to the part being produced. The user recognizes that the edges of every other coil will be opposite and arranges to uncoil either over or under the coil so the edge condition entering the press is always the same (Figure 2–23). It is unusual for edge shape to be of great concern, but when it is, the tool designer or press room supervisor will want to understand what to expect so it can be taken into account when planning the part fabrication. When heavy gauge, soft, narrow metal is required, concavity across the strip and some edge distortion are unavoidable in slit strip.

Coil set, the curvature which remains in a strip when it is unwound from a coil, is an inherent characteristic. The degree of this coil set is dependent on a number of factors. The final coiling operation takes place after slitting, and some measure of control over coil set can be

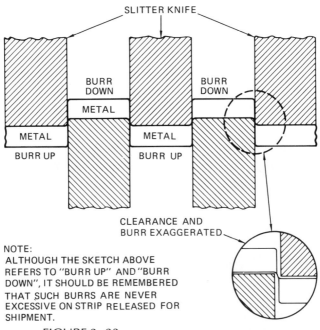

FIGURE 2–23
Burr up-burr down relationship in slitting setup.

exercised at this process stage. However, there are frequently other considerations that also have a bearing. For annealed tempers and the lightly cold-rolled tempers such as 1/4 Hard and 1/2 Hard, coil set may be established during final coiling. The degree of set will be least when the largest inside diameter which is compatible with the specified gauge and weight can be used. For the harder rolled tempers and lighter gauges, the coil set is actually controlled in the final rolling operation, rather than during coiling, and is usually kept to a minimum.

If the specified outside coil diameter is small because the user lacks handling equipment for large coils, it is desirable that the inside diameter be proportionately reduced. A coil with only a few wraps of metal because of a large inside diameter is flimsy and almost impossible to pack so it will not be damaged in transit. Coil size perferences by the mill take into consideration the desirability of a firm coil suitable for safe shipment (Figures 2–24 and 2–25).

When the user feeds the metal to his press to fabricate parts, it is certainly desirable that they be flat and free of coil set. This is particularly true for any multiple-station press with progressive dies where the part is carried in a web. Coiled metal inherently has some degree of coil set, and this increases from the outer to the inner end of

FIGURE 2–24
Finished coils ready to be packed.

FIGURE 2–25
Finished coils, paper wrapped, and packed in boxes for protection during transit.

the coil. There is no practical way that the parts fabricator can be sure of feeding flat metal into the press from a coil other than to flatten it himself. Flattening rolls installed ahead of the press will quickly pay for themselves through the increased production rates possible with coiled metal compared to hand-feeding flat lengths.

As pointed out earlier in this chapter, various coppers and copper alloys are processed in different coil lengths, depending on how they must be cast and their amenability to hot rolling. From the standpoint of manufacturing productivity, both in the brass mill and in the user's fabrication processes, the longer the coil, the greater is the productivity. The user of coiled sheet or strip will find it to his advantage from a productivity standpoint to use the coil size which the mill processes as a full length coil or an even fraction of this. Coil size requirements are obviously important to both the producer and the user.

In general, coil lengths are not stated in feet but are more conveniently measured in terms of weight. Pounds per inch of width is the convenient unit used by brass mills and is expressed in increments of 100 lb. Readily hot-rollable materials such as the coppers and brasses yield mill coils weighing from 300 to 400 lb./in. of width. A 25 in. wide coil will weigh from 7500 to 10,000 lb. In this instance, it would be best for the user to order at least 8000 lb. of one size in full length coils. For example, an order for 8000 lbs. of metal 2 in. wide would allow 11 or 12 strips to be slit, depending on how much edge trim was taken by the mill. Each coil would weigh approximately 700 lbs. (2 in. wide $\times$ 350 lb./in. of width), and a total of 7700 or 8400 lb. would be realized, which would fill the order. If only 4000 lb. were required, the desirable situation for the mill would be to cut the full length coil in half. Each coil received by the user would weigh 350 lb. (2 in. wide $\times$ 175 lb/in. of width), and the total would be 3850 or 4200 lbs., depending on whether 11 or 12 strips were slit. The mill would have a half length coil 25 in. wide remaining, which hopefully could be processed further and applied on another order for 4000 lbs. or less. By allowing some flexibility in coil size requirements, the user helps to maintain productivity levels in the mill, and by using larger coils where possible, will improve productivity in his own plant.

The mill always ships the largest acceptable coil to the user, unless order size or physical limitations militate against it. For example, a 0.5 in. wide strip of metal, at 400 lb./in. of width, would provide a coil weighing 200 lb. However, the outside diameter would be large, and because of the narrow strip width it would be almost impossible to coil the metal tightly enough. Thin gauge, narrow widths may be traverse wound on reels so that long coils can be conveniently handled. In fact, when this is done, several coils can be joined by butt welding to provide single lengths of 500 or 1000 lb. per reel. The butt welded joints are dressed to provide gauge and width uniformity, and the joint is sound enough that it will commonly take the press forming. The part containing the joint is not normally usable.

Light gauge metal in wider widths is frequently tightly wound on heavy cardboard cores to protect it from damage by handling or transit. In other instances, cores may be inserted in the coils or wooden crosses may be used to protect against coil collapse. As noted earlier, flimsy coils can result from specifying a large inside diameter and a small coil weight or an outside diameter such that the coil lacks the necessary rigidity.

Processing operations which follow final slitting are occasionally required. *Blanking* is one such operation. Blanking of squares or rectangles is generally done by cutting to length. The metal is first flattened and then cut to length on a flying shear. If the tolerance on length is such that it cannot be achieved on the automatic cutting lines, the cut lengths are resheared by hand. When circular blanks are required, they are die cut on a press. The tolerances for the diameter of circular blanks are the same as those for slit metal of corresponding width.

Edge rolling is another process which may follow final slitting. Edge rolling can produce rolled square edges, rounded edges, rounded corners, or rolled full rounded edges. It can only be done on a limited range of gauge, width, and temper combinations. Properties and tolerances are generally the same as those for similar slit edge products. Since the production capability for such products is dependent on a combination of several variables, no general statement on availability can be made. Each separate item must be reviewed by the producer to determine if the specifications can be met.

Further processing, such as tinning (electro or hot dip), is available from some mills. These subjects are discussed in later chapters with respect to their application and use.

3

CLASSIFICATION SYSTEMS
FOR ALLOYS
AND TEMPERS

THE CDA SYSTEM FOR NUMBERING COPPER AND COPPER ALLOYS

The Copper Development Association, Inc. (CDA) numbering system has been the accepted alloy designation system used in North America for wrought and cast copper alloy products. This system has been in use by the U.S. Government, the American Society for Testing and Materials (ASTM), the Society of Automotive Engineers (SAE), and almost all producers of copper and copper alloy wrought and cast products in North America.

This system is not a specification. It is an orderly method of defining and identifying coppers and copper alloys. It eliminates previous designation conflicts, and at the same time provides a workable method of identification for the products.

New alloys may be added or obsolete alloys discontinued from these lists by request from anyone and approval from the administrating body. New designations are assigned if an alloy meets three criteria:

1. The complete chemical composition is disclosed.
2. The copper or copper alloy is in commercial use.
3. The composition does not fall within the limits of any designated composition already in the list.

A numbering system has been developed by SAE and ASTM. It is called the *Unified Numbering System (UNS)*. Basically it will utilize many nationally used numbering systems and will adapt them to the UNS system in a manner which will avoid the confusion caused when the same number is used by different industries to identify different alloys.

As opposed to the CDA three-digit system, the UNS utilizes a prefix letter; C for coppers and copper alloys, A for aluminum and its alloys, N for nickel and nickel alloys, and so on, followed by a five-digit number. The CDA designation for cartridge brass is CDA 260; under the UNS this becomes C26000. New alloys of very similar composition could become C26001, C26010, and so on. The system allows for the addition of many new alloys and more exact record keeping and cross referencing without changing it in the future. The CDA will continue to administer the copper and copper alloy numbering systems, adopting UNS format, so these numbers will be coordinated with those of other metal industries.

The only other system in wide use is the one used by the International Standards Organization (ISO) and the Latin American countries. This designation is based on the chemical composition of the alloy. For example, Alloy C26000, which is 70% copper and 30% zinc, would be written as CuZn 30. For alloys containing several elements the designation becomes quite long, for example, CuNi 10 Fe 1 Mn.

The wrought coppers and copper alloys are grouped by composition in the UNS and CDA numbering systems.

The first of these groups is the *Coppers*. The UNS designations for the coppers run from C10000 to C15599. To be included in this group, the metal must have a minimum copper content of 99.3%.

Next are the *High Copper Alloys*, whose UNS designations cover C15600 through C19599. They have copper contents of 96% or greater but less than 99.3%. Most alloys in this group contain additions of cadmium, beryllium, chromium, or iron to give them greater strength than the coppers, but without drastic reductions in conductivity. They are primarily used in applications where thermal or electrical conductivity as well as strength is necessary to the finished product.

1 The copper-zinc alloy group, known as *Brasses* (UNS designations C20500-C28299), is the most popular in terms of pounds used because of the versatility of the alloys. They consist of alloys of copper and zinc ranging from Alloy C20500 (97% Cu, 3% Zn) to Alloy C28000 (61% Cu, 39% Zn). This group combines ease of manufacture with fair electrical conductivity, excellent forming and drawing properties, and good strength. These alloys find uses in many applications such as coins, medallions, jewelry, bullet jackets, ornaments, hinges, pen caps, electrical sockets, radiators, locksets, cartridge cases, musical instruments, electrical connectors, terminals, springs, and lamp bases.

2 The second copper-zinc group is *Leaded Brasses* with a zinc content of 32 to 39%, to which about 1 to 3% lead has been added, with the balance

copper. The lead is disseminated in small particles throughout the alloy, giving excellent machining qualities and also improving ease of blanking, shearing, sawing, and milling. These qualities, combined with mechanical properties similar to those of other high zinc brasses, make these alloys excellent candidates for use as gears, plates, clock and watch parts, keys, and a multitude of machined parts. These alloys fall in the UNS C30000 series.

The next group of alloys is commonly referred to as *Tin Brasses*. These contain zinc with additions of 0.5 to 2% tin. Their particular value is good corrosion resistance combined with strength. Covered by the UNS C40000 series, they have a wide range of uses which include fuse clips, contact springs, thrust washers, bushings, chains, electric fan blades, weather strip, electrical terminals, and connectors.

By addition of small amounts of phosphorus to eliminate oxides, copper-tin alloys are produced which are the *Phosphor Bronzes*. These are direct descendants of the ancient bronzes. There are actually two groups in this series, covered by the UNS C50000 designations. The first group, *Phosphor Bronzes,* possesses excellent tensile properties, as well as great resiliency, fatigue strength, and corrosion resistance. These alloys are especially valuable where resistance to alternating or cyclic stress is required, as in springs, diaphragms, bellows, and contact springs. They are also used in bearing plates and a wide range of functional parts. The second group, *Leaded Phosphor Bronzes,* has many of the strength attributes of unleaded phosphor bronzes, but with the addition of lead become outstanding performers as sleeve bearings, bushings, and thrust washers in all kinds of engines and drive trains. Zinc may be added, as in C54400, to further enhance the strength and hardness.

The next group is comprised of the UNS C60000 series of alloys. This includes three alloy families, which are *Aluminum Bronzes, Silicon Bronzes,* and *Miscellaneous Copper-Zinc Alloys.* The aluminum bronzes consist of copper with 2 to 13% aluminum added. These alloys have good strength properties and good formability. The most common uses are for nuts, bolts, corrosion-resistant vessels, machine parts, protective sheathings, and fasteners. There are also high performance alloys included in this group, which are, like the nickel silvers and phosphor bronzes, used in relay springs and other electronic applications where current carrying capacity is less important than strength.

The silicon bronzes are silicon-copper alloys which possess properties suitable for all types of welding, in addition to excellent hot-forming and cold-working properties. Typical uses include hydraulic-pressure fittings, screws, bolts, cable clamps, nuts, pole line hardware, rivets, electrical conduits, heat exchanger tubes, welding rods, bearing plates, piston rings, and screen wire.

The third group in the C60000 series is copper-zinc alloys that may have additions of aluminum, cobalt, manganese, nickel, and others. These have a very wide range of properties and uses and should be reviewed individually for overall suitability for a particular application. They find their widest use in areas in which corrosion resistance and strength are important; for example, electrical equipment applications.

The last wrought alloy group is covered by the UNS designations in the C70000 series. Again there are three basic divisions: the *Cupro-Nickels,* the *Nickel Silvers,* and the *Leaded Nickel Silvers.* The cupro-nickels are copper-nickel alloys which have many attributes. They have good forming qualities, high strength, and excellent corrosion resistance. They are used in a wide variety of applications, including condenser tubes, electronics and communications equipment, coins and medallions, and jewelry.

The nickel silvers are copper-zinc alloys with additions of nickel. They possess many of the same qualities of tarnish resistance and oxidation resistance as cupro-nickels with the exception that the additions of zinc make them stronger but less resistant to stress corrosion. The high zinc alloys are used in many spring applications because of their strength and stiffness. Nickel silvers are also used in such applications as silver plated flatware, hollow ware, and musical instruments because of their silver color, good formability, and corrosion resistance.

The leaded nickel silvers are copper-nickel-zinc alloys with additions of lead. They have good blanking and machining properties and are used for key blanks, large coins and medallions, jewelry, and other, similar applications where good machining qualities and a silver color are desirable.

A list of wrought coppers and copper alloys and their compositions is given in the appendix.

TEMPER DESIGNATIONS

The copper and copper alloy industry defines *temper* as "the metallurgical structure and properties of a product resulting from thermal or mechanical processing treatments." This is an all-inclusive definition since, in any copper or copper alloy product, metallurgical structure is a major determinant of resistance to deformation during fabrication and use.

The structure of a cast product reflects the metal temperature at the time of pouring, the cooling rate, and the method of containment during solidification. The wrought structure of hot-worked products reflects the hot-working temperature and the amount of mechanical work imparted. The grain structure of cold-worked products reflects the original structure (cast or wrought) before cold working and the strain imparted by the cold work. The structure of annealed products reflects the structure prior to annealing as well as the annealing temperature and time. All of these characteristic microstructures can be readily observed in properly prepared specimens viewed with an optical metallurgical microscope.

Other structural changes which may result in measurable property changes are not so readily visible, but may be detected by an electron microscope or by X-ray techniques. In fact, any structural state which affects properties, whether observable or not, as long as it is controllable by thermal and mechanical treatments, is a temper. While some other metals industries refer to such states as "conditions," the copper and copper alloy industry prefers the more specific term "temper."

When used to describe a copper or copper alloy product, the temper name or designation defines a fundamental and characteristic state. Standard tempers have been established to describe a variety of such states. The three basic temper categories for single-phase copper and copper alloy plate, sheet, and strip are *hot-rolled, cold-rolled,* and *annealed.* Hot-rolled temper is generally only applicable to plate and is essentially the same as an annealed temper material having a similar grain structure. Sheet and strip are the products with which this book is concerned, and the annealed and cold-rolled tempers are most applicable to these products.

The annealed tempers of single-phase materials are determined by the grain size. Since properties such as tensile strength, yield strength, and elongation vary in a consistent manner with grain size, which is readily measurable in these metals, it is used as the temper criterion. In general, strength decreases and elongation increases as grain size increases. An exception is the case where the product is very thin and the grain size is large, so there are very few grains through the thickness. Here both strength and elongation decrease with increasing grain size. Hardness roughly parallels strength. However, meaningful hardness measurements require considerable knowledge, technique, and understanding on the part of the tester to apply the appropriate hardness test to the many alloys, tempers, and thicknesses of copper and copper alloy strip. Poorly chosen or conducted hardness tests are worse than none.

Each copper and copper alloy has a characteristic recrystallization and grain growth response to annealing. Alloy C26000, when rolled to a 40% or greater reduction in thickness, can be annealed to any of several grain size ranges. The grain size of copper alloy products is measured and reported in millimeters. (See Chapter 5.) The reported grain size is always an average grain diameter based on the grains observed by viewing a polished and etched surface of a specimen under a micro-scope. A representative specimen is used, and several estimates or grain size measurements are made and averaged. The grain size, as noted earlier, relates to the tensile properties in a consistent manner, and it is the property used as a control measure for most annealed copper alloy products.

In the case of coppers such as C10200, oxygen-free copper, or C11000, electrolytic tough pitch copper, and other oxygen-free or tough pitch coppers grain size is more difficult to determine. There is a strong tendency for preferred orientation in cold-rolled, annealed coppers. This often results in neighboring grains with the same orientation appearing to be one large grain, and the true grain size is difficult to measure. Therefore, minimum tensile strength and minimum percent elongation are often used as requirements for annealed coppers. Grain size in deoxidized coppers can be used as a temper specification in a manner similar to its use for single-phase copper alloys.

Some alloys can be annealed to produce very small average grain sizes. Alloy C26000, cartridge brass, for example, can be annealed to a large number of ranges of average grain size, including very small ones. Since such small grains are difficult to measure, tensile strength is the preferred criterion for these tempers. The tensile strength ranges produced by annealing to these fine grain size ranges are similar to those produced by cold rolling to 1/4 Hard or 1/2 Hard temper, and these tempers are called "annealed-to-temper 1/4 Hard" and "anealed-to-temper 1/2 Hard." The fine grain size provides a smooth surface after forming, and the higher elongations and low yield strengths of annealed-to-temper 1/4 Hard and 1/2 Hard provide excellent formability. Excellent gauge uniformity is another benefit of these special tempers. (See Chapter 14.)

The grain size control that can be achieved is different for different alloys and is an inherent characteristic related to the composition. Some alloys will have either a very fine grain size or a very coarse grain size. These are usually marketed only in the fine-grained condition and are annealed to meet tensile strength requirements. Alloy C19400, covered by ASTM B465, and Alloy C68800, covered by ASTM B592, are alloys with this characteristic.

The precipitation heat-treatable alloys such as the copper-beryllium alloys are a special case. They, when supplied annealed, have been solution heat-treated to put the alloying elements into a supersaturated solution. Then, after cold rolling or forming, the metal can be precipitation heat-treated to strengthen it to the desired final temper. Solution heat-treated and all other tempers of these alloys in strip form are processed to meet tensile test requirements.

The cold-rolled tempers of copper and copper alloy sheet and strip contain considerably fewer variations than the annealed tempers. The basic criterion of cold-rolled tempers is tensile strength. In their production, the annealed metal is work hardened by cold rolling to meet a specified tensile strength range. The higher the tensile strength, the greater the amount of cold reduction required to meet it. Each copper and copper alloy is produced to a series of cold-rolled tempers with the appropriate tensile strength range for that material for each cold-rolled temper.

Annealed temper names have often been ambiguous, with different meanings for different people. The names applied to cold-rolled tempers are meaningful, and their meanings are universal. The common rolled temper names are listed below:

1/8 Hard
1/4 Hard
1/2 Hard
3/4 Hard
Hard
Extra Hard
Spring
Extra Spring
Special Spring
Ultra Spring
Super Spring

For a given alloy the application of one of these temper names requires that it be rolled to meet standard tensile strength requirements. Alloy C26000, for example, is produced in the following rolled tempers with the standard tensile strengths given below:

Temper Name	Tensile Strength	
	Ksi	MPa
Quarter Hard (1/4 Hard)	49–59	388–407
Half Hard (1/2 Hard)	57–67	393–462
Three-quarters Hard (3/4 Hard)	64–74	441–510
Hard	71–81	490–555
Extra Hard	83–92	572–634
Spring	91–100	627–689
Extra Spring	95–104	655–717

In order to eliminate some of the misunderstandings concerning the tempers of copper and copper alloy products, the Copper Development Association developed an alphanumeric code system for all the temper names used in the ASTM standard specifications covering copper and copper alloy products. This system was subsequently adopted by ASTM as Standard Recommended Practice B601. These temper codes will no doubt be adopted to general use in the future.

BIBLIOGRAPHY

1. *Annual Book of ASTM Standards Designation E-527-74, Standard Recommended Practice for Numbering Metals and Alloys,* American Society for Testing and Materials Committee on Standards, November 1974.
2. "Application Data Sheet—Standard Designations for Copper and Copper Alloys," No. 101/75, Copper Development Association, Inc.
3. "The Mill Products of Olin Brass," Olin Corporation, Revised September 1974.

4

METALLURGY OF COPPER AND COPPER ALLOYS

ATOMIC AND CRYSTAL STRUCTURE

This chapter provides the foundation for understanding the metallurgy of copper and its alloys, their crystal structure, and their mechanical and physical properties. Copper occupies the 29th position in the periodic table of elements. In the Bohr-Sommerfeld theory of atomic structure, the number 29 identifies the number of electrons (negatively charged particles) around the nucleus. The nucleus is positively charged and is surrounded by four distinct electron shells. The fourth or outermost shell contains only one electron, positioned farthest from the positively charged nucleus so that it is the most loosely held. This is the valence electron.

Silver and gold are quite similar in atomic structure to copper, since each has only one electron in the outer shell. These three elements are grouped together in the periodic chart. All three metals have some unusual and useful properties in common. All are noted for their high thermal and electrical conductivities, properties of importance to the electrical and electronics designer. All are quite ductile, an attribute most important for ease of fabrication. Each is noted for nobility or corrosion resistance, a quality of utmost importance to those who concern themselves with product life. Another unusual and valuable characteristic is that copper, gold, and many of their alloys exhibit selective spectral reflectance. They are known for their bright red and golden colors. This is a major consideration for decorative applications where their rich, warm hues add aesthetic charm.

In the molten state copper atoms have no long-range, orderly arrangement. When molten copper or a copper alloy is cooled below its solidification temperature, the atoms become arranged in an orderly three-dimensional pattern. This pattern can be visualized as being made up of individual cubic unit cells, with one atom at each corner and one centered in each face. Figure 4–1 pictures this unit cell construction with centers of the atoms depicted as small spheres.

When molten metal begins to freeze, nuclei made up of many thousands of unit cells form and begin to grow by the addition of more unit cells whose axes are oriented in the same direction. Such an aggregate of unit cells with the same crystal orientation becomes a grain when solidification is complete. The number, size, and orientation of grains in a cast structure are dependent on the rate and direction of heat extraction. Figures 4–2A, B, and C illustrate the progressive growth of cast grains in the molten metal under conditions of nondirectional heat extraction. Random orientation of crystalline grain structures occurs under these conditions, and Figure 4–2D illustrates cast grains of uniform size and random orientation.

In copper, as in all metals, the atoms are held together by *metallic bonds*. These bonds embrace large aggregates of atoms. Each valence electron loses its association with any given atom and is free to move about in the solid. These free electrons form a negatively charged electron cloud that surrounds the array of positively charged nuclei making up the crystal structure. The attraction between the nuclei and

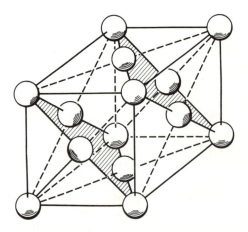

FIGURE 4–1

Model of face-centered cubic structure of copper, showing unit cell with atoms at the eight corners and at the centers of the six faces. Two close-packed planes are shown by shading. (Richman [8], p. 38)

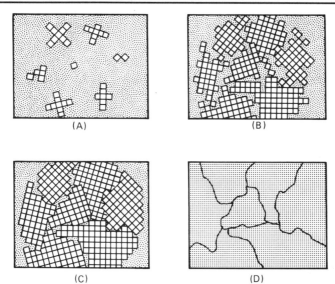

FIGURE 4-2
Representation of crystallization of molten metal by nucleation and grain growth: *A*, nucleation; *B*, grain growth; *C*, completion of crystallization; *D*, grain boundaries. (Allen [1], p. 80)

the electron cloud forms the metallic bond. Electrical and thermal conductivities are based upon the high mobility of electrons in the cloud; these electrons are free to move through the metal. A small potential produces a large current (defined by electron flow), which in turn means good conductivity. The high electrical conductivity of copper is one of this metal's assets.

PLASTIC DEFORMATION AND SLIP

Malleability, the quality of being able to be plastically deformed without fracture, is a characteristic of metals. Copper has this property of ductility in more than the usual degree.

Plastic deformation in a single crystal or grain occurs by the combined motion of many thin layers or sheets of atoms sliding over one another much like cards in a deck, as depicted in Figure 4-3. The surface or plane on which this slippage develops is normally the one where the atoms are closest together. In *face-centered cubic lattices* this close-packed plane cuts through three corners of each cube, as shown by the shading in Figure 4-1. These are the *primary slip planes,* and the geometry of the

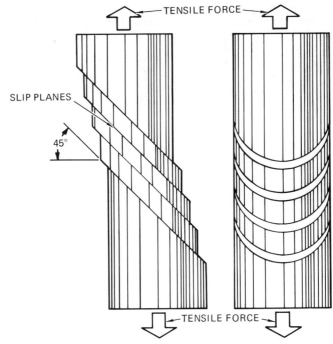

FIGURE 4–3

Orientation of slip planes at 45° to direction of stress. (Allen [1], p. 87)

cube structure provides four sets of them. In addition, it can be shown that for each set of slip planes there exist three directions in which it is possible for the planes to move. Therefore there are 12 possibilities for slip. This explains why copper and its alloys with this crystalline structure are well suited for applications requiring superior ductility. Some copper alloys, as will be noted later, have crystal structures which are *body-centered cubic*. This structure has fewer available slip planes, and these alloys are less ductile.

For loads applied in a single direction, maximum shear stress occurs at an angle of 45° to the direction of the applied load. The grains in which the slip planes and directions are most favorably oriented (45° to the direction of applied load) move or yield first, as indicated in Figure 4–3.

The calculated shear stress required to cause all the atoms in one plane simultaneously to slip over the atoms in the adjacent plane is extremely high. In fact, it is more than 1000 times higher than what is actually observed. This apparent paradox has been resolved by the discovery of atom *dislocations* or crystal imperfections in the grains. These allow the atomic forces to be overcome one atom at a time and thus greatly reduce the forces needed for deformation. Figure 4–4 illustrates one type of such defect, an edge dislocation. In the upper series of sketches, *A* represents a perfect crystal, *B* represents a crystal in which there is an edge dislocation, and *C* and *D* represent stress causing slip to take place and the vacancy caused by the dislocation to be filled; *E, F, G,* and *H* are the same as *A, B, C,* and *D* except that the dislocation is in a different location in the crystal. Because of these dislocations the bonding forces between the upper and lower halves of the crystal are weakened. Slip occurs easily, and the defect moves along the interface as shown. Dislocations develop during freezing. Many additional ones generated by plastic flow can and do interfere with one another and also pile up at structural boundaries. They act as barriers to further slip. Thus deforming copper at room temperature increases its strength and at the same time reduces its ductility.

EFFECTS OF ALLOYING ELEMENTS ON STRUCTURE AND PROPERTIES

Often, properties such as higher tensile strength or hardness, wear resistance, machinability, special corrosion resistance, or combinations of these, beyond what is available in unalloyed copper, are needed. To

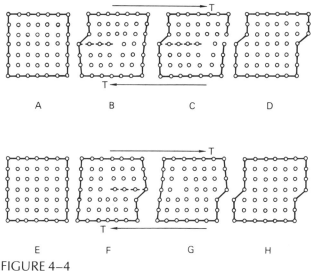

FIGURE 4–4
Generation and movement of an edge dislocation. (Taylor [10])

achieve such changes in properties, various amounts of another element such as nickel, lead, tin, and zinc are alloyed with copper.

If copper is melted and a small amount of another element, such as one of the above-mentioned, is added, it will dissolve to form a uniform, homogeneous liquid solution. When frozen, in practically every case, the alloying metal will remain in solution to form a solid solution. The added element becomes an integral part of the copper crystal. In such *solid-solution alloys*, the face-centered cubic structure of copper is usually maintained. Atoms of the added element replace copper atoms in the same lattice positions, forming a substitutional solid solution. Basically, such *substitutional* solid solutions are likely if the atoms of the alloy element are approximately the same size as the copper atoms.

There is also another type called an interstitial solid solution. In this case, the solute atoms are squeezed in between the solvent atoms at *interstitial* positions. Generally, this involves alloy element atoms much smaller than copper. This type of copper alloy does not have much commercial importance.

A *phase* is a homogeneous aggregation of matter which is distinct from neighboring phases. It possesses different properties and is separated from its neighbors by a perceptible boundary. Each phase of an alloy system is designated by a letter from the Greek alphabet. Copper-zinc alloys containing zinc in quantities less than 38%, under equilibrium conditions at room temperature, will be single-phase, *alpha* (α), alloys with face-centered cubic crystal structures. Copper-zinc alloys containing from 38 to approximately 45% zinc will be a mixture of two phases— alpha (α) face-centered cubic and *beta* (β) body-centered cubic crystal structure. Each phase has different properties and behaves differently in terms of workability. Both have good hot-working properties. The alpha phase is also ductile and readily worked at room temperatures, but the beta phase is harder and more brittle and has poor workability at room temperature. Figure 4–5 depicts the unit cell construction of the beta phase.

In this type of unit cell, an atom is found in the center and at each corner of the cube, giving rise to the term *body-centered cubic*.

Single-phase solid-solution alloys with face-centered cubic crystals behave much like unalloyed copper. Since their basic unit cell is essentially the same, we would expect the working characteristics to be generally similar. Some changes do occur, however, since the substitution of foreign elements tends to increase the strength and hardness and decrease conductivity.

The increase in strength relates back to greater resistance to slip or to the movement of dislocations. The solute atoms, because of their slightly different size, distort the lattice structure, interfering with slip. As a

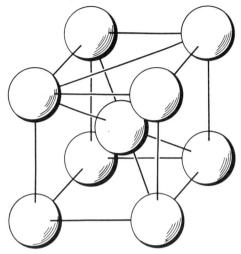

FIGURE 4–5
Model of the body-centered cubic structure of beta brass, showing unit cell with atoms at eight corners and one in the center of the cube body. (Richoan [8], p. 35)

dislocation moves along the slip plane, when it encounters the solute atom and the distorted lattice, extra energy is required for it to proceed. This *solid-solution strengthening* effect of alloying elements does not necessarily result in a loss of ductility. In fact, in many alloys such as the copper-zinc alloys the strength and ductility both increase with increasing zinc content up to the level where the beta phase begins to be present, as is shown later in Figure 4–8.

Electrical and thermal conductivity are decreased by the presence of foreign atoms. An in-depth discussion of exactly why this happens is rather complex; it suffices to note that resistance to electron flow is increased because the uniformity of the pure copper lattice is disturbed. As the number and severity of the imperfections increase the conductivity decreases.

PHASE DIAGRAMS AND PRECIPITATION HARDENING

A convenient way to show what happens to the metallurgical structure and hence the properties of the copper alloy system as progressive amounts of an alloying element are added is by means of a *phase* or *constitution diagram*. Such a diagram shows what phase or phases will be found in the alloy's structure at any selected percent of alloy element present and at any chosen temperature, from room temperature to above the melting point. In these diagrams, it is assumed that the alloy has been held at the particular temperature long enough for any structural change to be completed and that the system has arrived at equilibrium.

The simplest kind of phase diagram results when the alloying element is completely soluble in all percentages and forms no new phases at any level of composition or temperature. Copper-nickel alloys comprise such a system, as shown in Figure 4–6. Nickel has unlimited solid solubility in copper. Like copper, nickel has a face-centered cubic crystal structure. It freely substitutes in the copper lattice, because the atomic radius of nickel is approximately equal to that of copper.

When an alloying element has an atomic radius appreciably different from that of copper, substitution of its atoms for copper atoms in the face-centered cubic lattice is limited. In the case of zinc, as can be observed in Figure 4–7, a maximum of about 32% of zinc can be accommodated by the alpha face-centered cubic structure at 903°C and approximately 39% at 200°C. When more than these amounts of zinc are present, other phases are formed upon solidification. The beta phase has a body-centered cubic crystalline structure. Copper-zinc alloys which are mixtures of the alpha and beta phases have some commercial value as wrought products; however the beta phase is quite brittle, and cold working becomes increasingly difficult as the amount of beta present increases. Copper-aluminum alloys go through a similar transition at about 9.4% aluminum. Lead, which differs widely from copper in atomic radius, is virtually insoluble and separates from the alloy matrix.

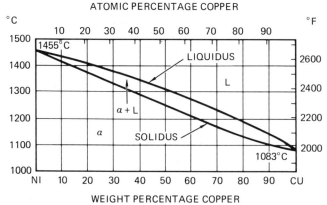

FIGURE 4–6
The copper-nickel system. (Pilling and Kihlgren [7-, p. 1198)

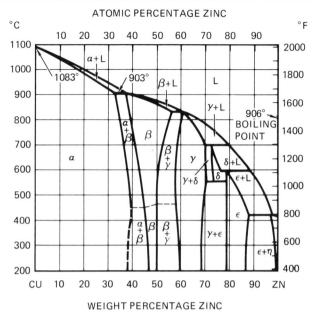

FIGURE 4–7
The copper-zinc system. (Phillips and Brick [6], p. 1206)

Separation of lead in this fashion is what makes leaded brass a free-machining alloy. The particles of lead not only act as a solid lubricant between metal and tool but also prevent tool wear by causing a discontinuous chip to be produced.

The influence of the beta phase on the strength and ductility of high zinc brasses is evident in the test data shown in Figure 4–8. The percent elongation drops rapidly as the beta present increases. The tensile strength peaks and then also drops rapidly.

Cooling during normal processing is usually so rapid that the equilibrium conditions on which the phase diagram is based are not achieved. A brass containing 35% zinc cooled to room temperature at the normal rate is likely to contain some beta phase. By heating it and holding a temperature of 650°C, which is observed in the phase diagram to be in the alpha phase for that composition, the zinc diffuses and the beta converts to alpha. This transformation is desirable to provide the improved cold workability of the 100% alpha structure.

In some copper alloy systems, when the limit of solid solubility of the addition element is exceeded, the new phase which appears in the structure is an intermetallic compound, which may appear as a precipi-

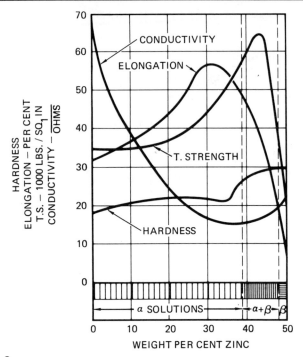

FIGURE 4–8
The effect of increasing zinc additions to copper on the tensile strength, hardness, elongation, and electrical conductivity. (Allen [1], p. 333)

tate. Figure 4–9 shows the high copper portion of the copper-beryllium phase diagram, which is the classic example of such a copper alloy system. The beta phase decomposes into alpha and gamma (γ) at 575°C. The gamma phase in the system corresponds to a form of the compound CuBe.

Beryllium coppers are precipitation-hardenable alloys. In Figure 4–9 the alloy containing 1% beryllium and 99% copper is observed to consist of two phases at temperatures of 500°C and lower. Upon heating the alloy for sufficient time at a temperature above 500°C, the beryllium is dissolved in the copper to form additional alpha. By cooling the alpha structure to room temperature rapidly by a cold water quench, formation of the gamma phase is suppressed. The alpha phase now consists of a supersaturated solution of beryllium in copper. Heating the solution-heat-treated alloy to a temperature of approximately 325°C for 2 to 3 hours will cause the gamma phase to precipitate out in a finely dispersed form. A dispersion of gamma precipitate so fine that it cannot be seen

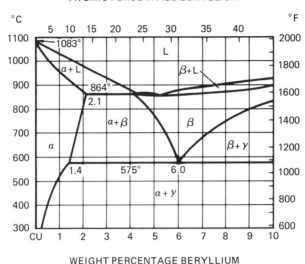

ATOMIC PERCENTAGE BERYLLIUM

WEIGHT PERCENTAGE BERYLLIUM

FIGURE 4-9
The copper-beryllium system. (Silliman [9], p. 1176)

under a light microscope provides maximum strengthening. The gamma particles act as barriers to dislocation movement in all directions.

Careful control is exercised during this precipitation-hardening process, because there is a level of reheating temperature and time when maximum strength will be reached. This strength level is directly related to the size of the precipitated particles. At higher temperatures or longer times, particle size increases and strength decreases. For high strength, small particles which are bound very tightly to the alpha structure and which retain a great deal of matrix identity are needed. These particles cause a widely felt stress field within the atomic structure of the matrix. This, in turn, presents a formidable barrier to dislocation movement. The precipitation-hardenable beryllium-copper alloys offer the highest strength of all the common copper alloys when appropriately cold worked and heat treated.

The presence of a second phase does not always mean an increase in strength and hardness. Leaded alloys, for instance, contain a lead-rich second phase. The large size of the particles and their nature with respect to the matrix create a situation in which hardening does not occur. The hardening and strengthening influence of any precipitated second phase depends on particle size and its manner of occurrence in the matrix structure. There are other elements which occur as precipitated phases in copper alloys, but none has a strengthening effect to the degree that beryllium does.

Up to this point we have been concerned with copper and *binary* (two-element) alloy systems. If the properties of a pure metal can be manipulated by the addition of one alloying element, even greater manipulation may result by adding several elements. In fact, this has been accomplished with a great deal of success. When three or four elements are involved, we refer to the system as a *ternary* (three-element) or a *quaternary* (four-element) system. These systems also can be either single phase or multiphase. When in a solidified alloy the alloying elements are soluble in copper at the existing temperature and concentration, the phase or phases present are solid solutions, and the strengthening mechanism is solid-solution strengthening. When the alloying elements form very fine precipitated intermetallic compounds, the strengthening mechanism is precipitation hardening. When the alloying elements form coarse particles of intermetallic compounds in the matrix, dispersion strengthening occurs. These are the strengthening mechanism which are commonly found in copper alloys in the form of sheet and strip.

EFFECT OF IMPURITIES

Impurities are elements present in a metal which are not added or retained intentionally. They may be present in copper because they occur with the element in its natural state, or are picked up during the refining process, or are inadvertently added along with recycled scrap melted as part of the furnace charge. Impurities are generally controlled to such levels as to be innocuous. If present in harmful amounts, they interfere with processing of the product or may alter the physical or mechanical properties. The chemical analysis requirements of each copper or copper alloy include limits for allowable impurities. Generally specifications will list the impurities which are likely to be present and known to be detrimental to processing, or to alter the physical or mechanical properties.

Impurity concentrations are usually very low compared to the level of intentionally added alloying elements. The sum of impurity elements such as As, Sb, Bi, Sn, Pb, Fe, Ni, Se, Te, S, and O that are found in electrolytic tough pitch copper is less than 0.1%.

The electrical conductivity of copper is adversely affected by the presence of any foreign elements. Those that are soluble in the solid state lower the conductivity most. Some impurities analyzed to be present are in the form of stable oxides which they form in the presence of the oxygen dissolved in tough pitch copper. In this form the element is not in solution in the copper, and only the portion dissolved in the copper lowers the conductivity. Examples of this type of impurity are iron and lead. Other elements, such as antimony and cadmium, form oxides which are unstable and may decompose during cooling of the solidified copper at temperatures above 700°C and be partially effective in lowering the conductivity of tough pitch copper. Finally, there are some impurity elements which have less affinity for oxygen than does copper. These do not form oxides at low concentration levels, even when oxygen is present in the metal. Nickel, for example, remains in solution in the copper, and the presence of oxygen does not influence its effect on conductivity.

The conductivity of copper is lowered by the presence of impurities. Figure 4–10 shows the effect of each of several impurity elements on the electrical conductivity of tough pitch copper. Since annealed tough pitch copper is required to have a minimum electrical conductivity of 100% IACS, the impurity levels are quite low.

It is a rare occurrence when impurity levels in copper and copper alloy sheet or strip are high enough to have any effect on its application. The major problem, when it occurs, is in the brass mill. Since copper and copper alloys are virtually indestructible in the normal course of events, they are recycled over and over. A constant vigilance is practiced to be sure that scrap for recycling does not contain unknown mixtures.

A not uncommon occurrence is the mixing of yellow brass and leaded yellow brass scrap by carelessness in a fabricator's plant where both are used. Since as little as 0.03% lead in brass can cause the metal to fracture in hot rolling, there is little likelihood that brass with higher lead

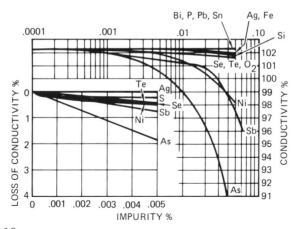

FIGURE 4–10
Decrease of the conductivity of tough pitch copper with increasing content of impurities. (Butts [2])

impurity would ever be shipped from a brass mill with hot-rolling equipment. The same reasoning applies to all hot-rollable coppers and copper alloys with respect to lead and other impurities which cause hot shortness (the condition of having a weak or partially molten structure at hot-rolling temperatures).

Iron is probably the most common impurity found in copper alloys because so much of the equipment used in metal production is steel. Tramp steel in the form of nuts, bolts, and so on can easily be dropped from equipment and get into scrap or raw materials. As little as 0.03% iron, particularly if a minute quantity of phosphorus is also present, can raise the annealing temperature and restrict grain growth in brass. This situation would interrupt production in the brass mill to the extent that, even if the impurities were not disclosed by analysis, they would be discovered during subsequent processing.

The illustrations above are indicative of how deleterious impurities cause such difficulty during mill processing that it is unlikely that the strip user will ever receive metal with sufficient impurities to cause problems. This is not true of all metal and has not always been true of the copper alloys, but the production methods and controls required to meet the quality and economic demands of the market have brought about this situation for copper.

BIBLIOGRAPHY

1. Allen, S. K., *Metallurgy, Theory and Practice,* American Technical Society, 1969.
2. Butts, A., *Copper,* Reinhold Publishing Corp., New York, 1954.
3. Finn, R. A., *Copper, Brass and Bronze Castings—Their Structures, Properties and Applications,* NonFerrous Founders' Society, Cleveland, Ohio, 1963.
4. Guy, A. E., *Elements of Physical Metallurgy,* Addison-Wesley Publishing Co., Reading, Mass., 1959.
5. Koehler, J. S., F. Seitz, W. T. Read, Jr., W. Shockley, and E. Orowan, *Dislocations in Metals,* Institute of Metals Division A.I.M.M.E., 1954.
6. Phillips, A. and R. M. Brick, *Dislocation in Metals,* Institute of Metals Division A.I.M.M.E., 1954.
7. Pilling, N. B., and T. E. Kihlgren, *Metals Handbook,* American Society for Metals, 1948.
8. Richman, Marc H., *An Introduction to the Science of Metals,* Blaisdell Publishing Co., 1967.
9. Silliman, H. F., *An Introduction to the Science of Metals,* Blaisdell Publishing Co., 1967.
10. Taylor, G. I., *Proc. Roy. Soc. (London),* **A145,** 362, 1934.
11. Van Vlack, L. H. *Elements of Material Science,* Addison-Wesley Publishing Co., Reading, Mass., 1964.
12. Wilkins, R. A., and E. S. Bunn, *Copper and Copper-Base Alloys,* McGraw-Hill Book Co., New York, 1943.

5

MECHANICAL PROPERTIES AND GRAIN SIZE

The attributes of copper and copper alloys which are of chief interest to design engineers are their mechanical, physical, and chemical properties. *Mechanical properties* such as tensile strength and yield strength relate to the way the metal strains or deforms under stress at various temperatures. They depend to a large extent on crystal or grain structure and its condition. *Physical properties* such as electrical and thermal conductivity, or coefficient of expansion, are conferred primarily by the basic electron configuration. The copper alloy systems have physical properties based on copper; however the physical and mechanical properties of alloys are different from those of copper because of the influence of the alloying elements on the basic crystal structure. The amount of difference between the properties of copper and those of any copper alloy system depends on the amount of the alloying element that is present. For example, Alloy C22000 (90 copper-10 zinc) Brass is mechanically and physically different from Alloy C26000 (70 copper-30 zinc) brass, although they are both copper-zinc alloys, because the amount of zinc present is different.

While physical properties are usually considered to be constant and inherent to the alloy, thermal treatments or mechanical working of some alloys can sometimes affect them significantly. The single-phase wrought alloys do have substantially invariant physical properties, but in multiphase or precipitation-hardening alloys, because of changes in amounts and kinds of phases resulting from different methods of processing, physical properties, such as electrical conductivity, can be drastically altered.

MECHANICAL PROPERTIES

A structural basis for mechanical properties is established by alloying. Cold deformation (work hardening) or annealing (softening) is used to vary them to suit the need. Heat treatments can also yield a range of mechanical properties in alloys designed to be responsive to them. In mill-supplied material the mechanical properties of coppers and copper alloys are evaluated by one of the following tests:

Tension test
Grain size measurement

Hardness tests are sometimes used to approximate strength.

TENSION TEST

Engineering tension test data are used by design engineers to provide basic data on the strength of materials and by materials engineers as acceptance test criteria in the specification of the temper of sheet and strip (refer to Chapter 14). In this test, a specimen is subjected to an increasing uniaxial tensile load, and at the same time the extension of the specimen within a given gauge length is measured. The instantaneous tensile load divided by the original specimen cross-sectional area is, by definition, the *engineering stress*. The instantaneous change in gauge length divided by the original gauge length of the specimen is the *engineering strain*. A plot of the two determinations over the period of initial load to specimen fracture yields an engineering *stress-strain diagram*, such as is shown in Figure 5–1.

Stress is given in pounds per square inch (psi) of cross-sectional area or in units of thousands (ksi). In The International System of Units (SI), stress is expressed in megapascals (MPa), equivalent to meganewtons per square meter (MN/m^2). The conversion factor is: 1 $lbf/in.^2$ (1 psi) = 6.8948 kN/m^2 = 6.8948 kPa; 1000 $lbf/in.^2$ = 6.8948 MN/m^2 = 6.8948 MPa. Strain is stated in inches per inch (in./in.) or meters/meter (m/m). It is a dimensionless quantity and may be expressed as a percentage value.

The primary information usually obtained for copper alloys from a stress-strain curve includes:

Elastic modulus
Yield strength

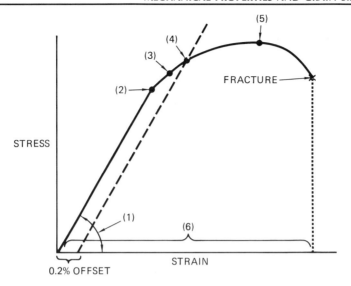

FIGURE 5–1
Sketch of tensile stress-strain diagram.

The proportional limit and elastic limit are other values which are sometimes estimated.

The *elastic modulus*, (1) Figure 5–1, also called *Young's modulus (E)*, is the ratio of stress to strain in the elastic portion of the curve and is based on *Hooke's law*, that is, at loads below the proportional limit, strain is a linear function of stress. It is a measure of stiffness. By definition, the *proportional limit*, (2) Figure 5–1, is the point within the elastic range where the stress-strain curve departs linearity. Again, by definition, the *elastic limit*, (3) Figure 5–1, is the maximum point on the curve where, if all load is removed, the specimen will return to its original length. No permanent set (strain) will remain. Beyond that point the specimen will show permanent deformation on load removal because of plastic strain.

The true proportional limit, although a desirable value to have, can be determined only by an exacting experimental technique using highly precise equipment for measuring strain. It is not usually available for use in engineering design because the true value is so difficult to determine. It is identified here to illustrate that there are two distinct regions in the curve: one where the metal is behaving elastically and the relationship is linear (Hookean), and the second where both elastic and plastic strain are present and the relationship is nonlinear. In some instances the linear part of the curve is very short, and the proportional limit and elastic limit differ greatly in strain and stress; then Young's modulus may not be determined accurately.

Cold deformation strengthens material, and also raises the elastic limit and the proportional limit. This increase in the useful elastic range is greater in some alloys than in others. When the effect is pronounced, the use of the conventional Young's modulus is not accurate for precise design. Steps can be taken to "correct" the modulus to provide a more accurate estimate of expected deflection under load for all stresses greater than the proportional limit.

A corrected modulus is the *secant modulus* (E_s), and it is used in lieu of Young's modulus in design calculations. The curves schematically drawn in Figure 5-2 illustrate the elastic modulus and elastic limit behavior for increasing cold work and the definition of secant modulus for design calculation work. The stress-strain curves for which the secant moduli lines E_{s3}, E_{s2}, and E_{s1} are drawn represent smaller amounts of cold work. That for which E is shown represents a large amount of cold work. When design stress exceeds the proportional limit or the elastic limit, secant moduli should be used in calculating deflection. Points P_1, P_2, P_3,

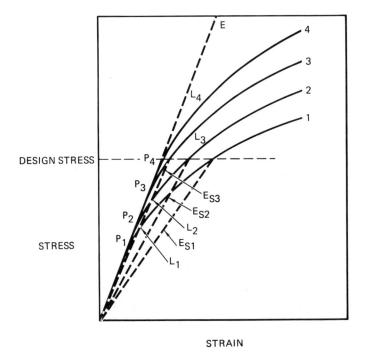

FIGURE 5-2
Effect of cold work on proportional limit, elastic limit, and secant modulus values.

and P_4 represent increasing proportional limits with increasing cold work. Points L_4, L_3, and L_2 illustrate how elastic limit may increase more than proportional limit with increasing amounts of cold work. E_{s3}, E_{s2}, and E_{s1} are secant moduli for a given "design stress." This design stress is shown to exceed the proportional limit for all tempers of the alloy except the most highly cold-worked temper, 4. For temper 4, Young's modulus could be used for calculating deflection. For tempers 3, 2, and 1, more accurate deflections could be calculated by using secant moduli E_{s3}, E_{s2}, and E_{s1}, respectively. Additional information on the determination and use of the secant modulus for design calculations is given in Chapter 13 on designing springs.

Yield strength, (4) Figure 5–1, for copper and copper alloys is an arbitrary level of strength used by designers. It has been standardized in the nonferrous industry as the stress on the stress-strain curve where the strain departs 0.2% from the modulus line. This particular value of yield strength is determined by drawing a line parallel to the elastic (Hookean) modulus but offset by 0.2% of strain. The point of intersection of this line with the stress-strain curve gives the yield strength at 0.2% offset, (4) Figure 5–1. The stress-strain curve is plotted by an attachment to the tension test. Other yield strength values which are often quoted for nonferrous metals are the 0.1% offset yield strength and the yield strength determined from the stress at 0.5% extension under load. All these yield strengths and the methods used to determine them are described in ASTM Method E8, Tension Testing of Metallic Materials.

Tensile strength, (5) Figure 5–1, is the maximum stress the material will sustain with uniform elongation. At that stress the onset of *necking* occurs (severe local contraction and elongation of the test specimen) and continues until the specimen fractures, generally at the point of necking. Because of the method of constructing the engineering stress-strain curve (use of original cross-sectional area) tensile strength (ultimate strength) is the maximum stress on the curve. The tensile strength is the maximum load divided by the original cross-sectional area of the specimen.

Elongation, (6) Figure 5–1, is reported as a percent increase in the original sample gauge length and is always measured after the specimen fractures. This is done by fitting the broken halves of the sample together and measuring the extension that has occurred. Its determination is most readily accomplished by the use of gauge marks inscribed on the sample prior to test. The original distance between these marks is the *gauge length*. Since elongation is always reported as a percent, gauge length must be included in the report, for example, percent elongation in 2 in. Two inches is the standard gauge length for sheet and strip tensile specimens.

The tension testing of metallic materials has been standardized in the United States by the American Society for Testing and Materials (ASTM).[1] These standards prescribe test equipment, sample preparation, test method, and method of reporting results. Standardization is extremely important to both user and producer. It provides a standardized procedure for determining properties for use in material specifications, and it provides a referee method in case of a disparity between test results on the same material tested in different laboratories. With careful specimen preparation and measurement, consistent results can be obtained.

GRAIN SIZE

As reviewed in Chapter 4 covering metallurgy, solid metallic materials are crystalline, that is, the element atoms, bound to make the bulk material, are arranged in a regular, repeating, three-dimensional array. It is possible in a pure metal to develop a single large crystal with no internal boundaries. Such crystals are, in fact, produced and used as engineering materials. By far, the bulk of metals are polycrystalline. They are composed of aggregations of crystals joined at boundaries. Aggregations of unit crystals are termed crystals or *grains*, and the boundaries between them are *grain boundaries*. In metals and single-phase alloys the unit crystal structure is identical in each grain. Grain boundaries are observed where there are different crystal orientations between adjacent grains. In two or more phase materials both the unit crystal structure and the crystal orientation may differ between grains.

The grain size of a metal or single-phase alloy is an important characteristic. As the average grain size decreases, the grain-boundary volume of the metal increases. As boundary volume increases, the material becomes more and more resistant to plastic flow, that is, it is stronger. Conversely, an increase in average grain size is accompanied by a reduction in strength.

Plastic flow occurs when the metal is stressed beyond the elastic limit. Plastic flow in metallic grains occurs by generation and subsequent movement of dislocations on parallel planes of high atomic density, as described in Chapter 4. Changes in crystal orientation from grain to grain impede dislocation movement. Slip direction must change when slip moves from one grain to the next. Additional stress is required. Therefore, the greater the number of grains, the stronger is the material. Grain size control has many ramifications in both strip processing and application, that is, in properties such as

[1]ASTM Standards E8 and E345.

Strength
Directionality
Formability
Surface appearance

Directionality in annealed metal occurs when metal is cold worked in the same direction between anneals, and more and more of the annealed grains become oriented in the same direction. The tensile properties are then different at 0, 45, and 90° to the working direction. Such a difference would be manifested in eared cups being produced from strip when blanked and cupped; see Figure 10–19. Directionality can be minimized if each successive process anneal is at a temperature which decreases the grain size from that of the preceding anneal.

Grain size measurement is predicated on the assumption that in fully annealed wrought metal the grain morphology is *equiaxed*, that is, the diameter of each grain is essentially the same for any axis of measurement. In reality, however, a metal structure is an aggregate of grains both in varying size and shape so that "grain size" is an estimate of the average grain diameter. Even if all grains were identical in size and shape, the specimen surface on which grain size is observed is a random plane cut through the structure. Hence the grains observed would appear to range in size from the maximum true diameter down to a size approaching zero.

Three methods of grain size estimation are used for coppers and copper alloys. These are described in ASTM Standard E112 and listed below in order of increasing accuracy:

Comparison procedure
Intercept (Heyn) procedure
Planimetric (Jeffries) procedure

For each method, sample preparation is the same, consisting of grinding and polishing a test specimen to a very smooth finish suitable for examination at high magnification. The polished surface is etched with a suitable chemical solution to reveal grain boundaries when it is examined under a metallurgical microscope.

The *comparison procedure* consists of matching the grain size of the specimen to one of a series of photomicrographs showing a range of grain sizes. Grain size charts are established for eact category of alloy to be measured, (i.e., copper or brass, twinned or untwinned, flat or contrast etch, etc.). A series of photomicrographs similar to those in

Figure 5–3 has been prepared by the ASTM for the full range of grain sizes specified for copper base alloys.

The Heyn intercept and Jeffries planimetric methods both involve actual grain count on the sample surface and, thus, are not as subject to operator bias as the comparative method.

In the *Heyn intercept method*, a line of known length is metallographically superimposed upon the sample surface and the number of grain-boundary intercepts with the line is counted. Several counts are made in different directions on the specimen. The length of the line in millimeters divided by the number of grains intersected by it gives the average grain diameter for one measurement, and the average of several such measurements is the average grain size. See Figure 2–17.

The *Jeffries planimetric procedure* is a grain area determination which is later reduced to the number of grains per square millimeter. This procedure involves a superimposition of a known area (5000 mm² circular or rectangular) upon an image of the specimen surface and the counting of whole grains within the area plus the grains intersected by the area boundary. The sum of whole grains plus one-half the number of grains intersected by the area boundary is multiplied by Jeffries multiplier f (variable with magnification) to yield the number of grains per square millimeter. This is converted to average grain diameter in millimeters through use of the conversion formula, $d = 1/\sqrt{m}$, where m is the number of grains per square millimeter.

HARDNESS TESTS

In hardness testing of metals, hardness is determined by pressing an indentor into the surface under a fixed load and measuring the depth or diameter of the resulting impression. The value obtained is a hardness number. These hardness numbers are relative measures of resistance to deformation and indicators of the strength, work hardening characteristics, and resistance to abrasion.

There are four widely recognized classes of hardness tests:

Brinell
Rockwell
Vickers
Knoop (micro-hardness)

Of these, the *Rockwell hardness test* and the *Rockwell superficial hardness test* are the most widely used in the United States copper and copper

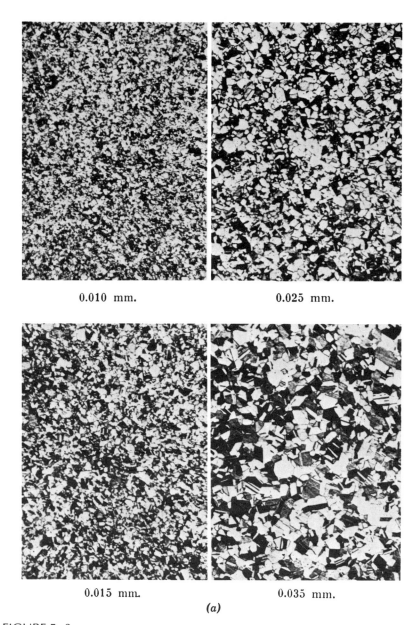

0.010 mm.   0.025 mm.

0.015 mm.   0.035 mm.

(a)

FIGURE 5–3

Standards for estimating grain size (nonferrous). American Society for Testing Materials Standard E2-36 (copyrighted) for annealed materials such as brass, bronze, and nickel silver. Average grain size is noted. Magnification of micrographs: 75 times.

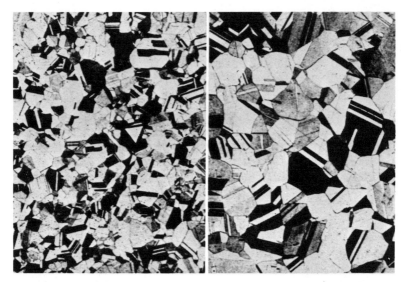

0.045 mm.   0.090 mm.

0.065 mm.   0.120 mm.

(b)

FIGURE 5–3   (continued)

74

alloy industry, primarily because of their applicability to a wide range of gauges, alloys, and tempers. These hardness tests are quick and easy to use, giving a direct reading with no supplementary measurements required. They are low in cost per test, and the test equipment is relatively inexpensive, easy to use, and easy to maintain. Therefore small laboratories often are limited to the use of hardness tests as indicators of temper, and hardness values are published in specifications to provide information which assists the user to make approximate temper judgments.

The *Brinell hardness test* is used primarily in the ferrous industry. The *Vickers* and *Knoop microhardness tests* are used more often in research or for checking small parts where a Rockwell test cannot be performed because the specimen is too small, or the section too thin.[1]

The standard Rockwell hardness scales are defined by the type and size of penetrator and the major penetrator load used in the test. A minor preload, used to establish the penetrator zero point, is constant for all scales at 10 kg. The major load applied, following the minor load, impresses the penetrator into the material surface to a depth indicative of specimen hardness. This distance or differential depth, (i.e., that from the zero point to full penetration) is atuomatically measured by the hardness machine and visually displayed as a hardness number on a direct reading dial. That number when recorded, together with the scale used, defines the specimen's hardness.

At the present time there are 15 standard Rockwell hardness scales and 6 superficial Rockwell hardness scales.

A schematic of the Rockwell principle of operation is shown in Figure 5–4.

The Rockwell superficial test is similar to the standard Rockwell test, except that a smaller minor load is utilized (3-kg) in conjunction with lighter major loads, thus extending the use of hardness testing down to thinner gauges and to relatively soft metals.

The Rockwell hardness scales most generally used in copper and copper alloy testing the listed in Table 5–1 with their applicable lower material thickness limits. Table 5–2 shows some hardness scales and lower limits which are applicable to hardened steels and other very hard metals.

Hardness testing of metallic materials has been standardized with respect to procedures and reporting by ASTM Standards E10, E18, E92, E140, and E384.

Hardness measurements can be affected by localized conditions at the surface being tested. As a result, it is not always possible to correlate the

---

[1]The European nonferrous hardness standard is the VPN or Vickers indentation test.

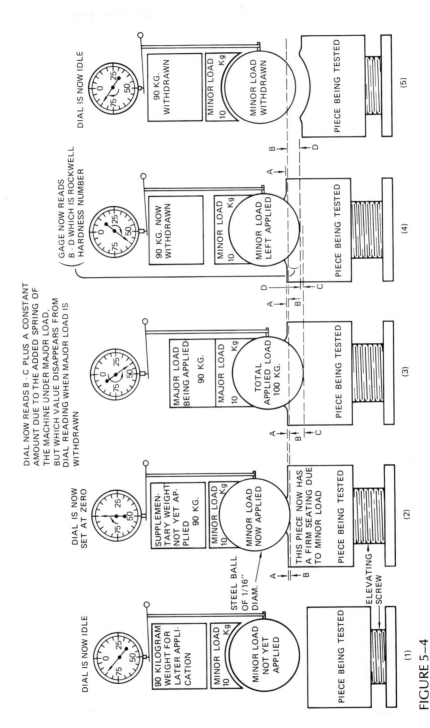

FIGURE 5—4

Illustrating schematically the principle of operation of the Rockwell hardness tester.

TABLE 5-1

Selector Chart for Scales Using $^1/_{16}$ In. Diameter Penetrator. Any greater thickness or hardness can be tested on the indicated scale.

Thickness In.	Rockwell Superficial Scale			Rockwell Scale	
	15T Dial Reading	30T Dial Reading	45T Dial Reading	F Dial Reading	B Dial Reading
0.010	91	. .	. .	. .	. .
0.012	86	. .	. .	. .	. .
0.014	81	79	. .	. .	. .
0.016	75	73	71	. .	. .
0.018	68	64	62	. .	. .
0.020	. .	55	53	. .	. .
0.022	. .	45	43	. .	. .
0.024	. .	34	31	98	94
0.026	. .	. .	18	91	87
0.028	. .	. .	4	85	80
0.030	. .	. .	. .	77	71
0.032	. .	. .	. .	69	62
0.034	. .	. .	. .	. .	52
0.036	. .	. .	. .	. .	40
0.038	. .	. .	. .	. .	28
0.040	. .	. .	. .	. .	. .

TABLE 5-2

Selector Chart for Scales Using Diamond Cone Penetrator. Any greater thickness or hardness can be tested on the indicated scale.

Thickness In.	Rockwell Superficial Scale			Rockwell Scale	
	15N Dial Reading	30N Dial Reading	45N Dial Reading	A Dial Reading	C Dial Reading
0.006	92	. .	. .	. .	. .
0.008	90	. .	. .	. .	. .
0.010	88	. .	. .	. .	. .
0.012	83	82	77	. .	. .
0.014	76	78.5	74	. .	. .
0.016	68	74	72	86	. .
0.018	. .	66	68	84	. .
0.020	. .	57	63	82	. .
0.022	. .	47	58	79	69
0.024	. .	. .	51	76	67
0.026	. .	. .	37	71	65
0.028	. .	. .	20	67	62
0.030	. .	. .	. .	60	57
0.032	. .	. .	. .	. .	52
0.034	. .	. .	. .	. .	45
0.036	. .	. .	. .	. .	37
0.038	. .	. .	. .	. .	28
0.040	. .	. .	. .	. .	20

hardness number obtained using one indentor, load, and hardness scale with that obtained using another, when testing a particular specimen. For each temper and thickness, there is a preferred choice of scale to provide the most reliable data.

While hardness test values can be indicative of the probable tensile strength of a material, correlations should be recognized as approximate only. They are limited to a given alloy composition, as illustrated in Figure 5–5, where the tensile strength and corresponding Rockwell 30T hardness number of brass Alloys C24000 and C26000 are plotted at progressively higher tempers. Note that there can be a significant difference in tensile strength when the hardness numbers are the same. Thus, where as a hardness number of 72 indicates a tensile strength of 80,000 psi for Extra Hard temper of Alloy C24000, the same hardness number, 72, indicates a tensile strength of 75,000 psi, or Hard temper, for Alloy C26000.

These are reasons why hardness tests, although convenient, are not used as acceptance and rejection criteria in ASTM specifications covering copper and copper alloy sheet and strip. Hardness ranges are given for each temper, but they are given only as information. Hardness tests are less precise and reproducible than tensile tests, and therefore, the latter are chosen for acceptance and rejection in ASTM specifications, along with grain size values. Tensile tests include the full thickness of sheet and strip specimens, and grain measurements are made beneath the surface metal so they are more representative of the metal than are hardness tests.

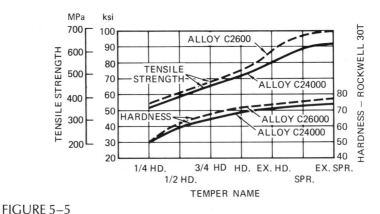

FIGURE 5–5

Tensile strength and Rockwell 30T hardness of Alloys C24000 and C26000 at increasing levels of temper. Note that similar hardness does not imply identical tensile strengths in the two alloys.

TENSILE STRENGTH AND TEMPER DESIGNATIONS

The use of temper designations in wrought copper alloys has evolved from the time when the material was used as either soft (annealed) or hard (cold worked). The designations have been continuously redefined and refined until temper designations are now quite precise and standardized within the industry. A temper code system developed by the industry has been adopted by ASTM as Recommended Practice B601-74. See Chapter 3.

The basic designations:

Annealed or Soft
1/4 Hard
1/2 Hard
3/4 Hard
Full Hard
Extra Hard
Spring
Extra Spring
Super Spring

are not all applicable to all alloys because of variations in rolling response. However, when rolled temper designations are used, an appropriate cold reduction from the annealed condition is required to achieve the temper or tensile strength.

Because absolute material control, that is, annealing to an exact grain size, rolling an exact reduction, or control to an exact chemistry, is not metallurgically possible or economically feasible, the rolled temper tensile strength will be within a range established by production history and practical limits of control. The tensile strength band is illustrated in the schematic shown in Figure 5–6, where the Alloy C26000 rolled temper designations are plotted against the strength ranges obtained by standard manufacturing techniques.

Annealed temper designations require either tensile properties or grain size requirements to be met, depending upon the alloy family. Tensile strength rather than grain size is specified for annealed tempers for some alloys which are inherently fine grained and from which grain size does not vary appreciably over a wide range of annealing temperatures. Also, the microstructure does not show sufficient grain contrast when etched to be accurately measured. Copper C11000 is an example where mechanical properties are not changed appreciably by annealing at temperatures well above the recrystallization temperature. Here,

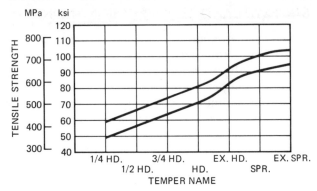

FIGURE 5–6
Upper and lower tensile strength limits for various rolled tempers of Alloy C26000.

minimum tensile strength and elongation requirements are specified to indicate the degree of annealing required. For some high copper alloys containing precipitated second phases, such as Alloy C19400, a tensile strength range is designated. The annealing temperature is chosen to also affect some physical properties such as electrical conductivity. The annealed tensile strength serves to reflect the proper temperature for treatment for this purpose as well as for the mechanical properties.

In the alloy families consisting of the brasses, tin brasses, phosphor bronzes, and nickel silvers, the common method of specification for annealed temper is by grain size. The strength and the formability of these metals are grain size dependent. Except in the very small grain sizes, relatively small changes in tensile strength and elongation result from quite different grain sizes in a particular alloy. Since grain size can be readily measured, it can be used to control annealing practices more accurately than mechanical property measurements.

Long experience has established the optimum grain size for the best response of each alloy in many different forming operations, such as spinning, cupping, and deep drawing. A further factor in determining the best grain size for an application is that the surface of parts formed may be rough ("orange peel") or smooth, depending on the grain size. When formed parts are to be polished and buffed after forming, grain size control is important for control of surface roughness. See Chapters 10 and 12 for further information on this.

In summary, the manufacture and application of such single-phase copper alloys in annealed tempers is most readily controlled by using grain size as the criterion for performance. The ASTM specifications covering these alloys, B36, B103, B121, B122, and B591, all list grain size requirements for them.

BIBLIOGRAPHY

1. Dieter, George E. Jr. *Mechanical Metallurgy*, McGraw-Hill Book Co., New York, 1961.
2. Specifications E8, E10, E18, E92, E112, E140, E345 and E348, American Society for Testing and Materials, Philadelphia Pa.

6

PHYSICAL PROPERTIES— ELECTRICAL CONDUCTIVITY, THERMAL CONDUCTIVITY, ELEVATED TEMPERATURE EFFECTS, AND DENSITY

The physical properties of copper and copper alloy sheet and strip which are most commonly considered by parts designers and materials engineers are the electrical and thermal properties. This chapter reviews the principles governing electrical and thermal conductivity in the materials. Since elevated temperatures are often encountered in applications where these properties are important, temperature effects are given. A brief discussion of density is included because it is a commonly used physical property in engineering metal requirements. The basic model is that of classical physics, which views atoms as made up of particles, in contrast to quantum mechanics, which describes them as mathematical wave-form equations. While the latter describes some properties more precisely, the former is easier to visualize.

ELECTRICAL CONDUCTIVITY

With technology so reliant on electrical energy, more than 75% of all copper and copper alloy products find their way into electrical or electronic applications. This is largely due to their inherently high *electrical conductivity*.

In 1913, values of electrical conductivity were established and expressed as a percent of a standard. The standard chosen was an annealed copper wire having a density of 8.89 g/cm³, a length of 1 m, a weight of 1 gr, and a resistance of 0.15328 Ω. The value of 100% IACS (*International Annealed Copper Standard*) at 20° (68°F) was assigned with a corresponding volume resistivity of 0.017241 Ω mm²/m. The percent IACS for any other material can then be calculated by

$$\% \text{ IACS} = \frac{0.017241\ \Omega\,\text{mm}^2/\text{m}}{\text{Volume Resistivity}} \times 100$$

There are many different units used in reporting resistivity and conductivity, and care should be used to choose the correct unit for use in this formula. Table 6–1 gives conversion factors for the units most commonly used. It is not uncommon for commercial copper products to have greater than 100% IACS conductivity. This has resulted from improved processing techniques developed in the 65 years since the standard was established.

In order to understand the effects temperature and impurity levels have on conductivity, it is necessary to have a model of the mechanism used in the conducting of electrical charges. Metals are interatomically bound by "metallic bonds," as noted in Chapter 4. Like valence electrons, the electrons from the outer shells of the atoms are free to move throughout the metal structure, acting like a negative electron cloud. The opposing charges of this electron cloud and the atomic nuclei provide the attractive force which bonds the metal atoms together.

The outer shell electrons not only are important in forming the "metallic bonds" but also give metal its ability to conduct electricity. Because they are not "anchored" to a specific atom, their own energy allows them to move within the crystal structure in any direction with equal velocity. With an electric field superimposed on the material, electrons moving towards the positive pole are given energy and accelerated. Any movement towards the negative pole will consume energy, resulting in a decreased velocity in that direction. The net effect is a movement of electrons towards the positive pole. Figure 6–1 illustrates this electron movement within metals. The greater this net movement, the higher is the electrical conductivity.

Metals such as copper whose outer shell electrons have great freedom have high electrical conductivity. The limiting factor on the net movement of the electrons is the number of deflections and reflections resulting from encounters with local electric fields around individual

TABLE 6-1
Conversion Formulas for Commonly-Used Units of Electrical Conductivity and Resistivity.

To obtain	Conductivity Units				Resistivity Units		
(Given number of ─ as a; Use this formula ─)	% IACS	megmhos/ centimeter	megmhos/ inch	mhos – ft/ circular mil	microhm – centimeters	microhm – inches	ohms – circular mil/ft
% IACS	a	$172.41a$	$67.88a$	$1037a$	$\dfrac{172.41}{a}$	$\dfrac{67.88}{a}$	$\dfrac{1037}{a}$
megmhos/centimeter	$.0058a$	a	$0.3937a$	$6.015a$	$\dfrac{1}{a}$	$\dfrac{0.3937}{a}$	$\dfrac{6.015}{a}$
megmhos/inch	$.01473a$	$2.540a$	a	$15.28a$	$\dfrac{2.540}{a}$	$\dfrac{1}{a}$	$\dfrac{15.28}{a}$
mhos-ft/circular mil	$.09643a$	$0.1662a$	$.06543a$	a	$\dfrac{0.1662}{a}$	$\dfrac{.06543}{a}$	$\dfrac{1}{a}$
microhm-centimeters	$\dfrac{172.41}{a}$	$\dfrac{1}{a}$	$\dfrac{2.540}{a}$	$\dfrac{0.1662}{a}$	a	$2.540a$	$0.1662a$
microhm-inches	$\dfrac{67.88}{a}$	$\dfrac{0.3937}{a}$	$\dfrac{1}{a}$	$\dfrac{.06543}{a}$	$0.3937a$	a	$.06543a$
ohms-circular mil/ft	$\dfrac{1037}{a}$	$\dfrac{6.015}{a}$	$\dfrac{15.28}{a}$	$\dfrac{1}{a}$	$6.015a$	$15.28a$	a

Rows grouped: "% IACS" through "mhos-ft/circular mil" = Conductivity Units; "microhm-centimeters" through "ohms-circular mil/ft" = Resistivity Units.

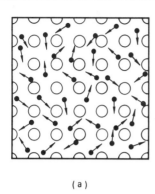
(a)

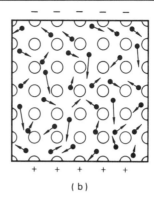
(b)

FIGURE 6-1

Electron movement in metals. (a) No electric field: Electron movement is random with equal velocities in all directions. (b) Superimposed electric field: Electrons moving towards the positive pole are accelerated, and those moving towards the negative pole are decelerated. Net movement is towards the positive pole.

atoms. The fewer of these, the longer is the *mean free path* between direction changes. This permits greater acceleration and deceleration to occur, increasing the net movement of electrons in the positive direction. The result is higher conductivity.

EFFECT OF ALLOY ADDITIONS AND IMPURITIES

Although elements such as nickel, iron, and silicon are deliberately added to improve strength or some other properties of copper, they must be considered *impurities* with respect to electrical conductivity. Impurities (including such alloy additions), provided they are soluble in copper, will drastically affect the electrical conductivity of copper. When dissolved, these impurities create discontinuities in the electrical field within the metal structure. The length of the mean free path of the copper's electrons is shortened by interference from the discontinuities introduced by the impurity, reducing electron velocity and lowering the electrical conductivity.

Electrical conductivity is affected significantly only when impurities (including alloying additions) are dissolved and go into solid solution. Figure 6-2 illustrates how impurities in solution disrupt the crystal lattice. Impurities which remain out of solution in the form of compounds or inclusions do not distort the basic crystalline structure in a manner which decreases conductivity. The greater the solubility of an impurity, the larger is its effect on conductivity.

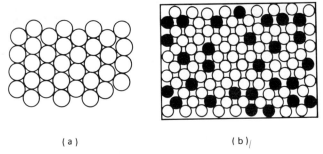

(a)   (b)

FIGURE 6-2

Effect of an alloying element on the crystal structure. (a) This is a graphic illustration of the crystal structure for pure copper. The uniform spacing between atoms results in uniform electric fields around each atom and within the entire structure. (b) With the addition of zinc, some copper atoms are replaced by zinc atoms in the crystal lattice to form a solid solution of zinc in copper. Zinc atoms, being of a different size, tend to distort the crystal structure and destroy the uniform electric fields. Electrons are reflected and deflected more often, decreasing the net movement towards the positive pole and decreasing electrical conductivity.

The effects of some alloying elements on the electrical conductivity of copper are seen in Figure 6–3. Elements such as silver, with one electron in its outer shell, and zinc, with only two, although both are soluble, have a relatively minor effect on conductivity when present in small quantities. Phosphorus, with five electrons in its outer shell, and nickel, with eight electrons, both also soluble in copper, greatly reduce conductivity, even when present in small amounts.

While the addition of soluble impurities decreases the conductivity, this effect can be lessened if the elements can be *precipitated* out of solution. Alloy C19400 (Cu—97.5%, Fe—2.35%, P—0.03%, Zn—0.12%) is one whose conductivity can be controlled by precipitation. During manufacture to finished gauge and temper, hot rolling and annealing are part of the processing operations. Hot-rolling temperatures are selected so that iron and phosphorus in the alloy are taken into solution, reducing the conductivity. However, with properly designed and controlled processing techniques, much of the iron and phosphorus can be precipitated out of solution by subsequent annealing, thus restoring conductivity. Since the process also has a strengthening effect, the result is an alloy with both good strength and conductivity.

The ability to retain strength and regain conductivity can be useful in applications requiring a brazing operation. An application such as is shown in Figure 6–4 requires the brazing of a contact to the end of the

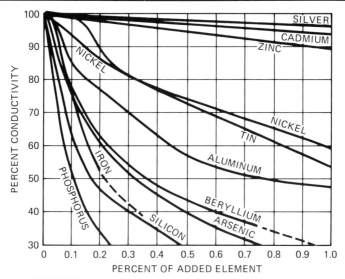

FIGURE 6-3

Effect of alloying elements on the conductivity of copper.

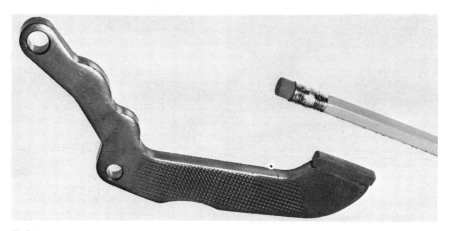

FIGURE 6-4

Example of a circuit-breaker arm where conductivity lost from brazing contacts in place was restored by postbrazing low temperature heat treatment.

heavy circuit breaker contact arm. Although conductivity throughout the entire arm may not be lost during this operation, the area neighboring the contact will probably become hot enough during brazing to cause some solution of the iron precipitate and a decrease in conductivity. By the use of a low-temperature-aging heat treatment subsequent to the brazing operation, conductivity can be restored without significant strength loss by reprecipitation of the copper-iron phosphides.

EFFECT OF TEMPERATURE

Heat is a form of energy and is absorbed by the atoms that make up the crystalline structure. As they absorb energy, the electrons become more agitated. With increased vibration about the atomic centers, the number of free electrons being reflected or deflected is greatly increased and conductivity is decreased. Conversely, as temperature is decreased below 20°C (68°F), conductivity increases for the opposite reason.

Specially refined and annealed coppers have had their conductivity checked at cryogenic temperatures, and it has been observed that the conductivity at very low temperature exceeded 1000% IACS. Figure 6–5 illustrates the relationship between electrical conductivity, resistivity, and temperature.

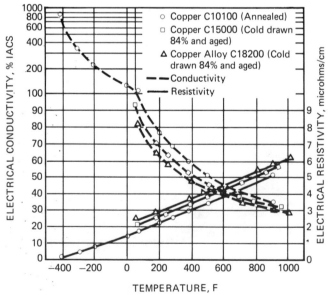

FIGURE 6–5

Electrical resistivity versus temperature.

THERMAL CONDUCTIVITY

Metals with high electrical conductivity usually have high thermal conductivity. This relationship is described by the *law of Wiedemann and Franz*:

$$\text{Constant} = \frac{\text{Thermal Conductivity}}{\text{Electrical Conductivity} \times \text{Temperature}}$$

The constant is 2.48×10^{-13} when cgs units are used. This relationship exists because the free electrons that conduct electricity by moving through the metal also carry over one-half the energy for thermal conduction. The remaining energy is transmitted by atomic vibration.

Since the atoms within a solid structure are held together by "bonding forces" (much like springs), any vibration by an atom is transmitted to its neighbor. Therefore, as one end is heated, this heat is absorbed by the surface atoms and through vibrations is transmitted through the entire structure. As with the flow of electrons, impurities impede this vibration, resulting in lowered thermal conductivity. Metals with high electrical conductivity normally have high thermal conductivity, also.

Next to electrical applications, applications involving the transfer of heat are the largest consumers of copper and copper alloys. Copper's ability to transfer heat rapidly and easily is critical to such products as automobile radiators and power-plant heat exchangers. Copper and copper alloys also are being used extensively in semiconductor lead frame applications as power requirements increase and the need for the removal of heat becomes more critical. Nickel and nickel-iron alloys, once the standard materials for semiconductor lead frames, are being replaced because copper materials have better corrosion resistance, higher electrical conductivity, and higher thermal conductivity. Where nickel and nickel-irons are still required, a copper pad might be used as a heat sink to act as a reservoir for the removal of excessive heat built up by the semiconductor.

Copper's high thermal conductivity is not always a benefit. Because heat is conducted away from the source so quickly, resistance welding of copper is difficult and is not recommended. With high thermal conducting copper, power requirements become so great that it is impracticable to weld in this manner. However, copper can be welded by the other methods discussed in Chapter 11.

Coppers with very high electrical conductivities show a decreased thermal conductivity as temperature is increased. However, thermal conductivity does not always decrease with increased temperature. When electrical conductivities are low at 20°C (68°F), they decrease far less per degree with increased temperature. Therefore, to satisfy the law of Wiedemann and Franz, thermal conductivity will increase.

Because of the disparate effects of temperature on the electrical and thermal conductivities, the designer should consider these possible effects on his product and its performance when choosing an alloy. In applications requiring both good electrical and thermal conductivities at elevated temperature, the alloy which will give the optimum performance, should be selected.

THERMAL EXPANSION

Because copper and its alloys are often used in applications such as electronic assemblies, where they are subjected to elevated temperatures by the assembly process or the service conditions, thermal expansion can influence selection. These materials have coefficients of thermal expansion which are higher than those of many other metals. This phenomenon is a result of atom vibrations. As heat is applied, the amplitudes of the vibrations increase, causing the average repulsive forces between atoms to increase relative to the attractive forces. This increases the mean distance between atoms, resulting in thermal expansion.

Thermal expansions bear a generally linear relationship to temperature. For some metals, such as the nickel alloys used for glass sealing, the shape of the expansion versus temperature curves is altered by alloying so they approximate the shape of the thermal expansion and contraction curves of the sealing glasses. Under these circumstances the materials can be joined since they expand and contract together during heating and cooling. There are some glasses which have thermal coefficients of expansion and contraction similar to those of copper and its alloys. Copper and copper alloys are sometimes used in glass sealing; in fact, excellent strengths can be provided by such seals. The joint actually occurs between the glass and the oxide film on the metal. For some coppers and copper alloys the oxides are extremely adherent, and very strong glass seals can be made. Because of the thermal expansion differences, most glass seals on copper are designed so that the metal contracts on the glass, and the stresses do not break the seal.

When silicon semiconductors are attached to copper alloy substrates, the attachment method takes into account the difference in the thermal coefficients of expansion and contraction between the copper and silicon. Nickel deposited on the copper alloy can serve as a buffer for both expansion and contraction and can prevent copper migration into the silicon when gold-silicon eutectic bonding is used. Also, conductive epoxy cements, which are flexible, have been used with these systems. The advantages of low cost, excellent corrosion resistance, high electrical

conductivity, high thermal conductivity, resistance to thermal softening, joinability, and excellent formability of the copper alloys used in such applications more than compensate for the need to provide for the mismatch in thermal coefficients.

In some applications a mismatch between the thermal coefficients of copper alloys and another metal is used to advantage. In some thermostat springs a copper alloy with a high thermal coefficient of expansion and contraction is laminated to another metal, such as a nickel alloy, with a low coefficient of thermal expansion. Such bimetal springs are activated by temperature change, as one metal expands or contracts more than the other.

DENSITY

Density is the *ratio of weight to unit of volume* of a substance. Copper and copper alloys have relatively high densities compared to other structural metals.

The densities of alloys are different from the densities of the pure metals composing them. For example, brass Alloy C26000 (70 Cu-30 Zn) has a lower density than copper C11000, even though heavier zinc atoms were added. The zinc more than offsets its higher atomic weight by greatly increasing the crystal structure volume. As a result, the weight per unit of volume of the alloy is less than that of either copper or zinc.

Density is an important factor of the metal cost of fabricated parts since it is a determinant of the weight of each piece. The cost of sheet and strip is based on weight, so lighter parts made from less dense metal will cost less. When materials perform equally well in an application, density may become the ultimate factor in choosing one over the other. Table 6–2 lists density values for various alloys. Chapter 15 discusses the economy of alloy selection, including the influence of density, and shows how to determine the net dollar effects on total costs of two alloys of different density.

TABLE 6-2
Density and Specific Gravity of Coppers and Copper Alloys

UNS Numbers	Density, lb/in	Specific Gravity, g/cc
C10200	0.323	8.94
C11000	0.322	8.92
C10920	0.322	8.92
C10930	0.322	8.92
C10940	0.322	8.92
C12200	0.323	8.94

TABLE 6-2 (Continued)

UNS Numbers	Density, lb/in	Specific Gravity, g/cc
C12900	0.321	8.90
C19400	0.317	8.77
C19500	0.322	8.92
C21000	0.320	8.86
C22000	0.318	8.80
C22600	0.317	8.77
C23000	0.316	8.75
C24000	0.313	8.67
C26000	0.308	8.53
C26200	0.308	8.53
C26800	0.306	8.47
C28000	0.303	8.39
C35000	0.306	8.47
C35300	0.306	8.47
C40500	0.319	8.83
C41100	0.318	8.80
C42200	0.318	8.80
C42500	0.317	8.77
C44300	0.308	8.53
C50500	0.321	8.89
C51000	0.320	8.86
C51100	0.320	8.86
C52100	0.318	8.80
C54400	0.321	8.89
C63800	0.299	8.28
C66400	0.317	8.77
C66700	0.308	8.53
C68800	0.296	8.20
C69910	0.277	7.67
C70400	0.323	8.94
C70500	0.323	8.94
C70600	0.323	8.94
C71000	0.323	8.94
C71300	0.323	8.94
C71500	0.323	8.94
C72500	0.321	8.89
C73500	0.319	8.83
C73800	0.312	8.65
C74000	0.314	8.69
C75200	0.316	8.73
C76100	0.317	8.77
C76200	0.317	8.77
C77000	0.314	8.70

BIBLIOGRAPHY

1. Kip, A. F., *Fundamentals of Electricity and Magnetism*, 2nd Ed., McGraw-Hill Book Co., New York, 1969.
2. Weidner, R. T., and R. L. Sells, *Elementary Classical Physics*, Vol. II, Allyn and Bacon, Boston, Mass., 1965.

7

OTHER ENGINEERING PROPERTIES—FATIGUE, CREEP, AND RELAXATION

In addition to the mechanical and physical properties usually considered when selecting a metal for a particular application, there are other engineering properties that need to be considered in relation to the design and application of the part. Of these, the less obvious properties of metals that are most often applicable to the behavior of mechanically functioning parts are fatigue, creep, and stress relaxation.

FATIGUE

Fatigue failures in metal parts are familiar to engineers and metallurgists who are regularly concerned with functional parts. They are sometimes attributed to the functional movement and sometimes to incidental vibration. An understanding of the phenomenon is desirable, but test data have not always been available to the metal user because fatigue tests are difficult to standardize and perform. The interplay of all the variables in fatigue phenomena is so complex that it is difficult to make a clear assement of a material's reaction in all circumstances. For sheet and strip, a bending load fatigue test is normally used because springs made from these products usually function in flexure.

In the most commonly used bending fatigue testing a standard sheet metal specimen is subjected to completely reversed bending. The bending action applies a known stress to the specimen first in one direction and then in the other. The number of cycles is counted up to

failure or to a selected maximum number without failure occurring. For copper and copper alloys 100,000,000 (10^8) cycles without failure is the usual end point of fatigue testing. The highest applied stress that can be applied without fatigue failure in 100,000,000 cycles is the "fatigue strength" of the material.

Fatigue properties of various metals tend to divide into two groups, those that have an endurance limit and those that do not. Many steels, ferrous metals, have endurance limits, but some grades and most nonferrous alloys do not exhibit endurance limits.

Fatigue test data are presented in the form of $S-N$ diagrams. Maximum bending stress is plotted against cycles to failure. By the use of $S-N$ curves, the number of cycles to failure can be determined if the applied stress is known, or the maximum safe bending stress can be determined from the number of cycles. A typical $S-N$ diagram is shown in Figure 7-1.

Two curves are illustrated in Figure 7-1, the first being representative of ferrous metals that have endurance limits, and the second being representative of nonferrous metals that have fatigue strengths rather than endurance limits. On the $S-N$ diagram for the ferrous material, the number of cycles to failure increases as the applied stress decreases until a minimum stress point or a "knee" on the curve is reached. Below this point failure will not occur regardless of the number of cycles imposed on the test sample. This point defines the *endurance limit* of the material.

The curve for the nonferrous material, on the other hand, does not exhibit an endurance limit. Rather, the number of cycles to failure

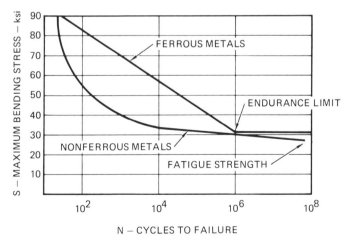

FIGURE 7-1
Typical form of S-N diagrams for ferrous and nonferrous metals.

continuously increases as the applied stress is decreased. Since an endurance limit cannot be defined, nonferrous fatigue properties are reported as fatigue strength at some particular number of cycles (usually 100,000,000 cycles, as noted above).

Fatigue failure in metals is thought to proceed from microcracks first generated at the surface by stress cycling. They propagate on a microscopic scale with each repeat cycle. Once the cracks have propagated to the point where the part cross section can no longer support the stresses imposed, catastrophic fracture occurs.

Part design is very important in avoiding fatigue failure. Notches normal to the imposed tensile stress are sources for stress concentration and, therefore, for premature crack generation. Tiny cracks, generated during the forming of a part because of improperly specified bend radii for the alloy and temper, can provide one immediate source. Rough edges on holes and burrs on blanked edges can also contribute to degradation of fatigue strength. These are important design and fabrication considerations in parts to be subjected to cyclic stressing. Designing parts so the maximum outer-fiber stress during operation will be below that necessary to result in fatigue failure is possible when fatigue strength data are available. Using such data, the materials engineer can choose alloys and tempers to meet the functional requirements of the part and avoid potential fatigue failure. Table 7–1 gives fatigue strength data for several copper alloys. The tabulated data are useful in comparing alloy, temper, directional, and heat treatment effects on fatigue strength. However, for design information, $S-N$ curves are required.

CREEP AND STRESS RUPTURE

In many engineering applications involving prolonged exposure at elevated temperatures, *creep* and *stress rupture* (or *creep rupture*) behavior are of major importance in alloy selection.

Creep is defined as the plastic deformation of a material as a function of time while under a constant stress. In conducting creep tests, a specimen similar to a tensile test specimen is placed in a furnace, heated uniformly to the test temperature, and then subjected to a constant load while the extension in length is carefully measured.

The shape of the plot of elongation versus time from a typical creep test will look like the curve in Figure 7–2. The creep curve can be divided into four stages:

1. Instantaneous strain. Elongation which occurs immediately upon application of load.

TABLE 7-1

Fatigue Strength Data for Coppers and Copper Alloys in Reverse Bend Testing—Tests Longitudinal[a] and 0.040 in. Thick Except when Otherwise Indicated

Alloy	Temper	Tensile Strength, ksi	0.2% Yield Strength, ksi	Fatigue Strength, ksi			Thicknesses, in., Other Than 0.040 in.	Direction of Specimen Axis[a]
				10^6 Cycles	10^7 Cycles	10^8 Cycles		
C19400	Annealed	47.3	21.9	>20	16.4	15.5		
	1/2 Hard	56.4	54.1	28.4	20.5	20.0		
	Hard	65.0	62.3	30.0	22.8	21.5		
	Extra Hard	66.4	63.2	—	21.7	19.6		45°
		63.6	58.5	—	21.9	17.0		
		68.4	63.9	23.5	20.9	20.0		Transverse
		66.1	62.9	22.7	21.8	21.7	0.060	
		67.3	63.9	—	21.5	18.8	0.020	
		67.6	65.3	18.7	17.8	17.5	0.010	
	Spring	71.2	67.9	—	21.9	21.8		
		69.5	65.0	—	22.0	21.8	0.060	
		72.9	69.9	—	23.1	21.0	0.020	
		69.7	67.7	19.1	18.2	17.5	0.010	
		70.3	67.9	27.5	21.4	19.0		
	Extra Spring	76.1	72.2	—	21.0	20.5		
	Super Spring	81.2	75.4	20.8	20.6	20.3	0.020	
		86.8	79.3	19.2	18.2	18.0	0.010	
		81.9	75.6	25.7	18.6	17.0	0.015	
		83.1	76.5	20.9	16.7	16.0	0.010	
C19500	Soft	61.8	40.1	28.0	25.8	25.5		
	Light Annealed	83.5	73.0	37.0	33.3	31.5		
	1/2 Hard	82.9	72.6	34.6	32.7	32.0		

TABLE 7-1 (Continued)

Alloy	Temper	Tensile Strength, ksi	0.2% Yield Strength, ksi	Fatigue Strength, ksi			Thicknesses, in., Other Than 0.040 in.	Direction of Specimen Axis[a]
				10^6 Cycles	10^7 Cycles	10^8 Cycles		
	Hard	96.1	91.6	36.2	32.2	31.5	0.030	
	Spring	98.0	89.3	39.0	35.0	34.0		
	Extra Spring	97.2	94.5	35.6	28.2	28.0		
C26000	1/2 Hard	61.0	36.0	26.5	21.4	21.0		
	Hard	80.3	70.6	—	24.5	23.5		
	Spring	91.2	83.8	29.1	27.0	26.6		
	Extra Spring	101.5	82.9	32.8	27.8	27.0		
		104.0	86.1	35.3	31.2	29.5		45°
		109.1	92.9	41.7	39.0	37.5		Transverse
	Super Spring	104.3	93.6	33.1	30.7	30.2		
C42500	Extra Hard	81.5	77.9	32.5	29.1	29.0		
	Extra Spring	90.5	86.2	36.5	36.0	36.0		
	Super Spring	102.8	92.7	39.5	33.0	32.0		
	Ultra Spring	108.2	96.5	43.6	36.9	33.0		
		103.3	96.2	39.7	37.7	37.5		
C51000	Hard	86.7	84.3	35.6	34.5	34.0		
	Spring	106.2	103.3	38.7	34.9	34.1		
	Super Spring	114.4	108.0	—	33.5	28.5		
	Ultra Spring	116.3	108.4	—	33.5	28.5		
C51000	Ultra Spring	119.7	113.6	34.2	32.9	32.5		
	H.T. 250°C[b]	113.7	110.0	46.8	41.7	39.5		
	H.T. 250°C[b]	133.5	130.0	58.8	58.1	58.0		Transverse

Alloy	Temper						Transverse
C52100	Hard	102.5	97.2	—	32.5	29.6	
	Spring	114.6	110.2	35.2	33.9	33.0	
	Super Spring	127.4	115.7	36.0	34.4	33.5	
	Ultra Spring	135.3	126.7	34.0	30.5	30.0	
		132.0	125.9	34.4	31.6	31.0	
C63800	Soft	78.8	52.0	36.9	34.4	34.0	
		85.8	59.2	27.3	26.6	26.5	
	1/4 Hard	88.3	66.7	33.5	32.1	32.0	
	3/4 Hard	109.1	91.8	38.3	35.2	35.0	
		115.7	98.8	36.5	34.5	34.5	
	Hard	117.0	102.0	40.6	39.6	39.4	0.050
	Spring	124.0	106.0	41.8	41.4	41.0	0.033
	Extra Spring	129.0	112.0	41.7	33.5	32.5	
	H.T. 300°C[b]	141.5	129.5	52.5	46.5	45.0	0.024
	H.T. 300°C[b]	172.5	155.5	59.9	57.7	57.5	0.024
C68800	Soft	80.5	56.5	33.0	23.0	22.0	
		84.2	58.5	35.8	32.4	32.0	
	1/4 Hard	93.2	78.3	32.9	32.1	32.0	
	Hard	115.0	97.0	37.7	27.2	26.0	
	Extra Hard	122.9	110.9	32.0	30.7	30.5	
	Spring	123.0	101.5	36.4	34.9	34.5	
		128.8	103.7	35.3	33.2	32.5	
C76200	3/4 Hard	95.6	93.8	36.4	33.4	33.0	
	Extra Hard	107.4	104.6	40.5	29.5	29.0	
	Spring	116.9	108.3	34.5	30.4	30.0	

[a]Longitudinal indicates the specimen's axis is parallel to the rolling direction; transverse indicates the specimen's axis is across the direction of rolling; 45° indicates the specimen's axis is at 45° to the direction of rolling.
[b]H.T.—Heat treated for 1 hour at temperature shown.

2. Primary creep. A period of decreasing strain rate due to work hardening from the initial load.
3. Secondary creep. A period of constant strain rate. Also called minimum creep rate.
4. Tertiary creep. A period of increasing strain rate, terminating in rupture.

The curve in Figure 7–2 illustrates these stages but is only a partial description of the performance because creep rate changes with temperature. A more complete illustration of a material's creep behavior is given in Figure 7–3, showing the secondary creep rate for a series of temperatures. For a given temperature, the secondary creep rate (%/hr) is shown as a function of applied stress (ksi). A log-log plot of the data results in a series of straight lines.

If the *creep strength* of a material is defined as the stress required to generate a specific creep rate (e.g., 10^{-5} %/hr), it can be plotted as a function of temperature, allowing a direct comparison of the creep performances of a series of alloys for which creep strengths have been determined for a series of temperatures.

Figure 7–3 gives secondary creep rates for Alloy C19500 at relatively high temperatures and small stresses. Another situation, which is probably more like that experienced in the application of copper alloy parts, such as electrical connectors, is that in which the stress is high and the temperature varies from room temperature to 125°C. Table 7–2 contains creep data based on stresses at 50% of 0.2% offset yield strength at a temperature of 125°C for several alloys and tempers. The creep rates

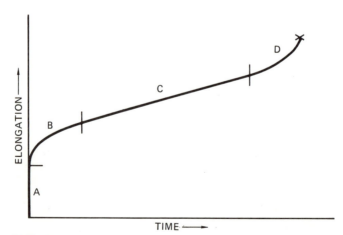

FIGURE 7–2
Typical creep-rupture curve displaying the four stages up to failure.

TABLE 7-2
Moderate Temperature Creep Data

Alloy	Temper	Tensile Strength, ksi	0.2% Yield Strength, ksi	Test Temperature, C°	Load (Stress), ksi	Load, % of Yield Strength	Total Creep, % in 2000 hr	Creep Rate, %/hr at 2000 hr × 10⁻⁵
C17200 (Be-Cu 25)	1/2 Hard	193.2	173.2	125	86.8	50	0.020	0.73
C17500 (Be-Cu 10)	Hard	110.2	103.0	125	51.5	50	0.032	1.33
C19400	Spring	71.2	67.9	125	34.0	50	0.055	0.31
C19500	1/2 Hard	90.3	88.0	125	44.0	50	0.104	0.95
	Spring	91.7	89.1	125	44.6	50	0.072	0.85
	Super Spring	99.6	94.4	125	47.2	50	0.077	0.81
C42500	3/4 Hard	69.8	65.8	125	32.9	50	0.093	2.07
	Spring	92.7	88.7	125	44.3	50	0.150	3.06
C51000	Hard	88.7	86.4	125	43.4	50	0.103	2.12
C63800	Spring	125.9	111.5	125	55.8	50	0.264	1.44
	Extra Spring	134.0	120.0	125	60.0	50	0.223	1.47
C66400	Hard	92.9	86.0	125	43.0	50	0.194	1.47
C68800	1/2 Hard	97.0	84.0	125	42.0	50	0.268	4.33
	Extra Hard	123.0	111.0	125	55.5	50	0.418	3.37
C72500	Hard	81.0	79.2	125	39.6	50	0.016	0.53
	Extra Hard	86.5	82.0	125	41.0	50	0.033	0.41
C76200	Spring	111.9	107.4	125	53.7	50	0.088	0.12

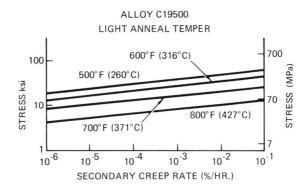

ALLOY C19500
LIGHT ANNEAL TEMPER

FIGURE 7–3
Secondary creep rates for Alloy C19500 at various temperatures.

after 2,000 hours are given in percent per hour. A value such as 0.307×10^{-5} %/hr might be interpreted to mean that, if the creep rate continued, it would give 0.0307% in 10,000 hours. The reason for the difference between this value and the 0.055% total creep in 2,000 hours is that the instantaneous strain is included in the 0.055% figure.

Stress-rupture tests measure the ability of a material to resist rupture at elevated temperatures. Although creep tests are quite helpful in evaluating performance at elevated temperatures, the time at which tertiary creep (and subsequent failure) will start can be just as important in some cases. The time for rupture at a given stress level, for example, is indicative of behavior at greater than designed-for loads.

For a given temperature, the time to failure is recorded as a function of the applied constant stress. A log-log plot of applied stresses versus time to failure for a series of temperatures will yield a series of linear relationships—one for each temperature. Figure 7–4 shows such a plot for Alloy C19500, which is one of the more creep resistant copper alloys. The discontinuity observed in the creep-rupture curve obtained at 800°F (427°C) is attributed to the dynamic recrystallization and associated formation of internal voids in specimens tested for over 100 hours.

The stress-rupture strength of a material may be defined as the stress to produce rupture at a specific time, that is, 1,000 or 100,000 hours. The relative performances of a series of alloys can be compared by comparing the rupture stresses for all at a given time and a series of temperatures. It is not likely that stress-rupture considerations will be necessary in sheet and strip applications, but they are frequently required in tube and plate applications.

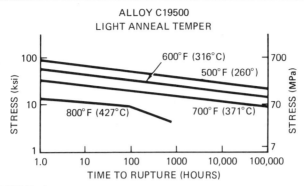

FIGURE 7–4
Typical stress-rupture curves for Alloy C19500 at various temperatures.

STRESS RELAXATION

Stress relaxation is a loss of stress while at constant strain. When a piece of metal is strained by application of an external stress, it reacts by developing an equal internal stress in the opposite direction to the applied stress. When the piece is locked into the strained position, the reactive stress relaxes, at first rapidly, and then more slowly, until it reaches a minimum level. The rate of stress relaxation and the minimum level reached depend on the alloy, temper, applied stress, temperature, and time.

One example of this is a bolt which has been tightened to apply a high compressive load holding two parts together. With the passage of time, and particularly at elevated temperatures, this compressive load decreases from its initial value. Another common example is a contact spring which is in a bent position to apply pressure to hold two contacts together. The load which the spring first exerts gradually relaxes at a rate dependent upon the initial load, the material from which the spring was made, and the temperature. When the application of the load stresses the spring beyond the proportional limit of the material, stress relaxation can be expected to occur. Figure 7–5 is a schematic stress relaxation curve showing how the stress in the strained material falls over a period of time.

Stress relaxation data for different alloys, tempers, temperatures, and directions relative to the rolling direction are being developed in research laboratories of some copper alloy producers, using methods described in ASTM STP 608. Data such as are shown in Tables 7–3 and 7–4 describe the materials tested. Plots of stress relaxation data such as those shown in Figures 7–6 and 7–7 are published to aid designers in

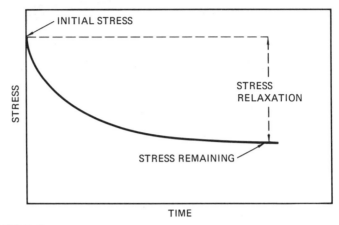

FIGURE 7–5
Schematic stress versus time curve defining the terms relating to stress relaxation.

TABLE 7-3
Room Temperature Tensile Properties of C51000 and C76200 Parallel to the Rolling Direction

Alloy	Young's Modulus × 10⁶ psi (× 10³ MN/m²)	0.2% Yield Strength, ksi (MN/m²)	Ultimate Tensile Strength, ksi (MN/m²)	Elongation in 2 in., %
C51000	16.3 (112.39)	110 (758)	111 (765)	2.6
C76200	18.3 (126.18)	105 (724)	107 (738)	2.3

TABLE 7-4
Stress Relaxation Data for C51000 and C76200 Parallel to the Rolling Direction

Alloy	Test Temperature, °C	Initial Stress, ksi (MN/m²)	Stress Remaining at 1000 Hours, ksi (MN/m²)
C51000	RT[a]	97 (669)	92 (634)
	105	95/90[b] (655/621)	72 (496)
C76200	RT	84 (579)	83 (572)
	105	83/81[b] (572/559)	73 (503)

[a] RT = room temperature.
[b] Stress remaining after 1 hour exposure at 105°C, that is, zero time stress.

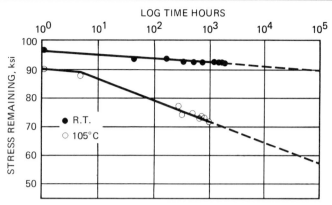

FIGURE 7-6
Stress remaining versus time for C51000. Multiply ksi by 6.895 to obtain MN/m².

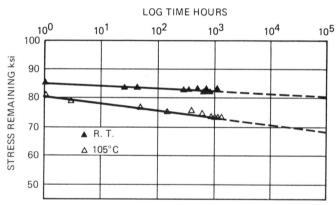

FIGURE 7-7
Stress remaining versus time for C76200. Multiply ksi by 6.895 to obtain MN/m².

materials selection. Such data are useful to engineers designing electrical spring contact arms, connectors, and terminals which depend on the spring characteristics and the gripping strength of the material to maintain good electrical contact.

In selecting an alloy-temper combination with acceptable stiffness and stress relaxation performance, adequate formability to make the part is also a consideration. The designer is confronted with the practical problem of part layout and the effect of orientation with respect to rolling direction on contact pressure and fabricability. In Chapter 13 on spring design considerations, these criteria are covered in more detail.

BIBLIOGRAPHY

1. ASTM Recommended Practice E328–72, Part C, for Stress Relaxation Tests for Materials and Structures.

2. Battelle Memorial Institute, *Prevention of the Failure of Metals Under Repeated Stress,* John Wiley & Sons, New York, 1941.

3. Cazaud, R., *Fatigue of Metals,* Chapman and Hall, London, 1953.

4. Fox, A., "The Effects of Extreme Cold Rolling on the Stress Relaxation Characteristics of CDA Copper Alloy 510 Strip," *J. Mater,* **6,** 422 (1971).

5. Garofalo, F., *Fundamentals of Creep and Creep-Rupture in Metals,* The Macmillan Co., New York, 1965.

6. Parikh, P., and E. Shapiro, "Stress Relaxation in Bending," ASTM STP–608, Recent Developments in Mechanical Testing, September 1976.

8

CORROSION RESISTANCE

The two general environments of most concern to engineers with respect to the corrosion resistance of coppers and copper alloys are the atmosphere and various aqueous media. While these are basically air and water, they are encountered with many gradations and variations in contamination with other chemicals. This chapter begins with a brief discussion of the general principles governing corrosion by atmospheres and aqueous solutions. This is followed by a review of engineering data from field experience on the performance of coppers and copper alloys in the two kinds of environments. Special types of corrosion, such as galvanic and biofouling corrosion, erosion-corrosion, and stress corrosion, are then discussed.

A. MECHANISMS OF CORROSION

1. Atmospheric Oxidation and Corrosion

Chemically clean surfaces of most engineering metals exposed to air react with the oxygen and oxidize quickly. The rate at which these oxides form is initially rapid but decreases as a protective film develops. These thin films impart corrosion resisting characteristics to copper and copper alloys. The oxide which forms on copper is largely cuprous oxide (Cu_2O). It may contain impurities such as sulfides and chlorides if these are present in the environment. Pilling and Bedworth have shown that for oxidation, if the ratio Md/mD (where M is the molecular weight of the oxide and D its density, m is the atomic weight of the metal multiplied by the number of metal atoms in the oxide formula, and d is the metal density) is greater than unity, the oxide film is protective; when less than unity it is nonprotective. (See Reference 3.) This ratio for cuprous oxide is 1.71.

The presence of alloying additions in copper such as nickel, zinc, tin, aluminum, and iron tends to reinforce the oxide film formed on copper; and in some cases (e.g., with nickel and aluminum) reduces the thickness of the steady-state film which is formed. Consequently, such alloys as C72500 (Cu-88.2%, Ni-9.5%, Sn-2.3%) have good oxidation resistance. They maintain good solderability during storage.

The oxides formed on copper alloys in the presence of moisture, either liquid, steam, or condensate, are similar to the oxides formed in air; but the rate of corrosion may be accelerated, particularly if sulfur-containing gases are also present. Atmospheric attack on copper and copper alloys is usually very uniform and occurs at a low rate. They can be expected to last for a long time with no protective coating.

Oxidation during processing or during service can be a source of problems. Oxides formed during processing, if excessive, can lead to galling of tools, decreased die life, uneven draws, and other difficulties during fabrication. Oxidation during service in electrical applications, due to corrosive environment or elevated temperature, can result in increased contact resistance and circuit failures. Process oxides develop during hot rolling and during annealing. For this reason, producers of copper and copper alloy sheet and strip usually mill the surface of their products to remove excess oxides prior to cold rolling. Annealing furnaces with controlled atmospheres are used to help keep oxidation to a minimum after the milling operation.

The removal of oxides from copper alloys is frequently necessary during processing and in preparing the surface for special coatings such as platings or solders. Hot (50 to 80°C) 10% sulfuric acid solution is frequently used to clean copper alloys, because it dissolves cupric oxide without attacking the metal. For alloys containing more than 85% copper, an oxidizing agent such as sodium dichromate is sometimes added to convert cuprous oxide to cupric oxide. Metal degreasing and cleaning equipment is commonly found in brass mills and is used to remove stains and oxide films that might be associated with annealing. A bright dip solution of sulfuric, nitric, and hydrochloric acids is frequently used to clean copper alloy air-annealed parts because of its fast, thorough action. Bright dips are usually used at room temperature. Table 8-1 lists some typical pickling and bright dip solutions.

In general, the higher the temperature, the greater is the amount of oxide formed on the surface. Consequently, it is the high temperature processing steps which must be controlled to prevent excessive oxidation. Surface condition also has an effect on the rate of oxidation in that a smooth, polished surface oxidizes more slowly than a rough surface.

TABLE 8-1
Cleaning Solutions for Coppers and Copper Alloys

Pickling Solutions	
1. Sulfuric acid pickle	
Sulfuric acid	1 gal
Water	9 gal
Temperature	52–80°C
2. Dichromate Pickle (for alloys over 85% Cu)	
Sulfuric acid	1 pint/gal
Sodium dichromate	2–4 oz/gal
Temperature	Room to 80°C
3. Ferric sulfate pickle (alloys over 85% Cu)	
Sulfuric acid	13 oz/gal
Ferric sulfate, anhydrous	13 oz/gal
Temperature	
Low copper alloys	50–60°C
High copper alloys	60–80°C
Bright Dip Solutions	
[a]1. Sulfuric acid	2 gal
Nitric acid	1 gal
Water	1.5 gal
Hydrochloric acid	0.5 fl oz
Temperature	Room
2. Phosphoric acid	55% by volume
Nitric acid	20%
Acetic acid	25%
Water	Keep below 10%
Temperature	55–80°C
[a]3. Sodium cyanide	4 oz/gal
35% Hydrogen peroxide	5 fl oz/gal
Temperature	Room

[a]These solutions should not be used by the uninitiated. Special precautions are required, and their use is not recommended without them. In addition, disposal of the solutions presents environmental hazards. These solutions are mentioned primarily as information.

Data source—see Reference 1.

2. Aqueous Corrosion

Corrosion at liquid-state temperatures generally proceeds by an *electrochemical mechanism*. Such corrosion involves a flow of electrons between anodes and cathodes on the metal surface through a conductive solution, the electrolyte. The anodic and cathodic areas may develop as the result of a variety of factors, such as inclusions, surface imperfections, differences in the orientation of grains, lack of chemical homogeneity, localized stresses, and external variations in environment. They are microscopic in size and so numerous as to be visually insepara- ble. A potential difference always exists between anode and cathode areas and results in the flow of current. Thus we have the basic elements of a corrosion cell: (1) anode; (2) cathode; (3) electrolyte; and (4) path for electron flow. These are shown in Figure 8–1, which is a schematic illustration of a corrosion cell.

Electrochemical action in the corrosion cell causes preferential dissolu- tion of the metal at the anode area. Positively charged metal ions detach themselves from the metal surface and enter the electrolyte, generating a counterflow of electrons. These electrons travel through the metal to the cathode area, where they take part in the cathodic reaction. Metal ions formed at the anode area, which is corroding, react directly with oxygen dissolved in the electrolyte to form a metal oxide or hydrated metal oxide corrosion product film. The anode and cathode areas on the metal surface continuously shift. This causes essentially uniform corro- sion to occur, and the corrosion products cover the metal surface evenly. If the anode remains at one point, localized corrosion such as pitting or dezincification may occur. The surface corrosion products resist current flow, slowing the anodic and cathodic reactions and reducing the rate of metal loss. The nature of the corrosion product film formed, and its resistance to further corrosion, are dependent on the metal or alloy corroding. Corrosion product films are more or less protective, depend- ing on their ionic and/or electronic resistance; therefore, some alloys have lower general corrosion rates than other alloys because their films are more protective. The resistance of the electrolyte can also affect

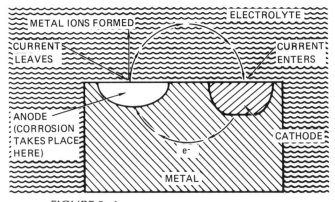

FIGURE 8–1
Schematic diagram of a corroding system or cell.

corrosion rates. For example, fresh water, which is less conductive than seawater, is also generally less corrosive.

The corrosion resistance of copper and its alloys is primarily due to the direct formation on the metal surface of a cuprous oxide film by the reactions

$$4CU + 2H_2O \rightleftharpoons 2CU_2O + 4H^+ + 4e^- \text{ (occurs at the anode)}$$
$$O_2 + 2H_2O + 4e^- \rightleftharpoons 4OH^- \text{ (occurs at the cathode)}$$

$$4Cu + O_2 \rightarrow 2\,Cu_2O \text{ (overall reaction)}$$

The cuprous oxide film formed during the corrosion of pure unalloyed copper is adherent to the metal surface; however, the structure of cuprous oxide is highly defective, being deficient in cations (Cu^+). Cuprous oxide is a P-type semiconductor in which each deficient positive charge is balanced by a single positive charge (or positive hole) in the vicinity of the cation vacancy, as shown in Figure 8–2. This is equivalent to a local valence change to the Cu^{2+} state. In such a defective structure, movement of both cations in the vacancy, and electrons, because of electronic change between Cu^+ and Cu^{2+}, is facilitated. Therefore, the anode and cathode reactions shown previously, being dependent on the movement of ions and electrons, are increased, thus increasing the corrosion rate.

Alloying additions, such as aluminum, zinc, tin, iron, and nickel, are frequently combined with copper to form the commonly used engineer-

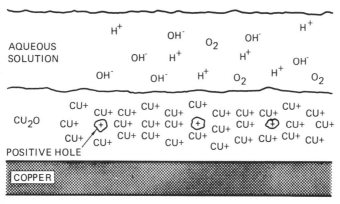

FIGURE 8–2

Illustration of defect nature of the cuprous oxide film formed during the corrosion of copper in an aqueous solutin. Alloying additions can plug these "holes," making the copper alloy more resistant to corrosion.

ing alloys of brass, bronze, nickel silver, and phosphor bronze. It has been found that these solid-solution alloying elements enter into the cuprous oxide corrosion product film, and may significantly reduce the corrosion rate (see North and Pryor [2]). This is accomplished by the alloying element plugging the positive holes in the corrosion product film, making it more difficult for ions to migrate to the oxide-corrodent interface and electrons to be conducted through the corrosion product film. The effectiveness of an alloying element in reducing the corrosion rate depends on a number of things, such as atom size (the atom must be small enough to fit into the vacancy), valence (the greater the valence, the more effective the alloying element will be in increasing the electronic resistance of the oxide film), and composition of the electrolyte. In general, the alloying additions listed above decrease the corrosion rate of copper.

B. FIELD EXPERIENCE

1. Atmospheric Oxidation and Corrosion

Atmospheric corrosion is of interest to the engineer because the parts he is designing, such as switches, relays, contacts, and other equipment, may be installed in areas of heavy industry where the air is polluted, or along coastal areas where the parts are subject to air containing salts from the ocean. Industrial pollutants such as sulfur dioxide, carbon dioxide, and oxides of nitrogen can form acids in the presence of moisture which can cause rapid corrosion of some materials. These polluted areas can also result in stress-corrosion cracking of susceptible alloy systems, as is discussed further along in this chapter.

The presence of moisture usually accelerates the rate of corrosion. The salty air near coastal areas contains chlorides which can cause dezincification, pitting, and stress-corrosion cracking of some copper alloys. Each of these will be discussed in more detail. Dezincification is quite prevalent in brasses containing 30% or more zinc and exposed to industrial atmospheres. Copper is often selected for its corrosion resistance in these types of environments. However, when severe service is expected, the alloys which possess the greatest resistance should be selected.

Table 8–2 is a listing of copper alloys and their resistance to corrosion in industrial and marine atmospheres. The data utilized in preparing this table were gathered by exposing flat plates of each material in typical industrial and coastal areas around the country. Groups of specimens were placed on racks on rooftops, and periodically some were removed and cleaned and the weight loss was determined. (The tests were conducted by Olin Corporation's Metal Research Laboratories.)

TABLE 8-2
Atmospheric Corrosion Data
Total Weight Loss in Milligrams per Square Centimeter (mg/cm²) and Instantaneous Corrosion Rate in Mils (0.001 in. penetration) per Year (Mpy).

Alloy	Temper	Location[a]	2 Years mg/cm²	2 Years Mpy	4 Years mg/cm²	4 Years Mpy
C11000	Hard	NH	1.93	0.043	3.76	0.041
		BR	3.45	0.076	6.66	0.074
		DB	4.55	0.087	8.00	0.088
		EA	1.88	0.042[b]	[c]	
C19400	Hard	NH	1.68	0.038	3.77	0.042
		BR	3.11	0.070	6.34	0.071
		DB	3.43	0.064	5.42	0.038
		EA	1.99	0.045[b]		
C19500	Spring	NH	1.80	0.040[b]		
		BR	2.15	0.047[b]		
		DB	4.44	0.098[b]		
		EA	3.67	0.081[b]		
C23000		NH	1.58	0.036	3.74	0.042
		BR	3.26	0.073[b]		
		DB	1.45	0.020	2.30	0.018
		EA	1.80	0.040[b]		
C26000	Hard	NH	2.24	0.052	4.56	0.053
		BR	4.35	0.093	7.64	0.044
		DB	0.66	0.014	1.15	0.005
		EA	1.70	0.039		
C42200	Hard	NH	1.77	0.040	4.20	0.047
		BR	3.56	0.080	7.50	0.084
		DB	2.35	0.045	3.54	0.029
		EA	1.68	0.038[b]		
C51000	Hard	NH	2.04	0.045	4.64	0.051
		BR	3.52	0.078	7.37	0.082
		DB	6.85	0.068	9.68	0.052
		EA	2.68	0.059[b]		
C52100	Hard	NH	1.22	0.027[b]		
		BR				
		DB	6.14	0.137[b]		
		EA	2.06	0.046[b]		
C63800	¾ Hard	NH	1.41	0.034[b]		
	Hard	BR	4.47	0.106[b]		
		DB	2.71	0.065[b]		
		EA	1.38	0.033[b]		
C66400	Hard	NH	1.60	0.036[b]		

TABLE 8-2 (Continued)

Alloy	Temper	Location[a]	2 Years mg/cm²	Mpy	4 Years mg/cm²	Mpy
		DB	2.60	0.058[b]		
		EA	2.82	0.063[b]		
C68800	½ Hard	NH	1.24	0.030[b]		
	Hard	BR	1.97	0.047[b]		
		DB	0.874	0.021[b]		
		EA	1.21	0.029[b]		
C77000	Extra	NH	2.36	0.053	4.75	0.108
	Hard	BR	4.54	0.089	8.62	0.108
		DB	0.606	0.014	1.36	0.014
		EA	1.68	0.038[b]		

[a] NH = New Haven, Conn.; BR = Brooklyn, N.Y.; DB = Daytona Beach, Fla.;
EA = East Alton, Ill.
[b] Value based on one data point.
[c] Blank = data not available.

The corrosion rates reported in this and subsequent tables are *instantaneous corrosion rates* at the time period indicated. Instantaneous corrosion rate is the slope of the tangent to the weight loss versus time curve at the time of interest and expresses the rate of change at that point. Most copper alloys exhibit weight loss-time curves like one of those in Figure 8-3. If the curve bends over with time, the instantaneous corrosion rate decreases. If the curve is linear, such as curve *A*, then the corrosion rate is constant.

The instantaneous corrosion rates can readily be calculated from weight-loss time curves like these. The tangential slope is used to determine weight loss per unit of surface area per unit of time (rate of change). Typical units for weight loss, surface area, and time are milligrams, square centimeters, and days, respectively. Observations in these units provide a corrosion rate in mg/cm²/day. It may be more useful to convert these units to inches of thickness per year or mils of thickness per year, using the conversion factors in Table 8–3. Density values can be found at the end of Chapter 6, Table 6-2.

2. Aqueous Corrosion

The general theory of film formation on copper alloys in aqueous environments has already been discussed. Copper alloys in general have excellent resistance to corrosion in aqueous environments and for

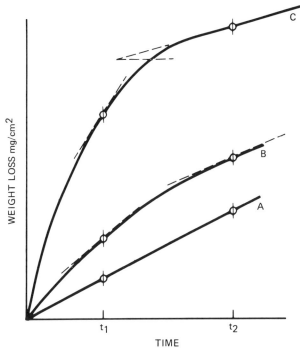

FIGURE 8–3

Schematic diagram showing various types of plots of corrosion weight loss versus time. Curve A—corrosion rate is the same at times t_1 and t_2 and is apparently at steady state at t_2. Curve B—Corrosion rate is decreasing, but has not established steady state at t_2. Curve C—corrosion rate is decreasing at t_1 and has established state at t_2.

centuries have been the material most often selected for service in these environments. Aqueous environments can generally be divided into three categories: (1) fresh water, which is generally characterized by low total dissolved solids, low chlorides, and relatively high percentage of scaling constituents, such as carbonates and sulfates; (2) brackish water, which is generally characterized by moderate to high total dissolved solids and chlorides in excess of 1,000 ppm (parts per million); (3) seawater, which contains high total dissolved solids (35,000 ppm) and high chlorides (19,000 ppm).

Historically, copper alloys have proved most resistant to the rigorous corrosion, biofouling, and crevice attack associated with seawater. The first metal hull sailing vessels were made from wood which was covered by a sheathing of copper. Later, copper and brass made up the major components of marine hardware and the heat exchange tubes needed

TABLE 8-3
Corrosion Rate Conversions

Multiply	By	To Obtain
mg per dm^2 per day (mdd)	0.00144/density	in. penetration per year (ipy)
mg per cm^2 per day	0.144/density	in. penetration per year (ipy)
mg per cm^2 per day	144/density	mils per year (mpy)
in. penetration per year (ipy)	696 × density	mg per dm^2 per day (mdd)

for steam powered sailing vessels. Brass was an excellent material for marine hardware such as pulleys, cranks, pumps, and other equipment. However, the severe corrosive action of seawater flowing inside heat exchanger tubes required newer materials. The brasses were alloyed with tin and aluminum initially to form tin brasses which more successfully resisted these corrosive environments. Later, nickel was alloyed with copper to create the copper-nickels which, with iron added, are still the most popular copper alloy choice for seawater service. Copper and brass tubes and pipe have long been used for carrying potable water in buildings and residential homes and for water heating systems. Table 8–4 shows the relative rankings of some copper alloys in a 3.4% brine solution cascading down a trough at approximately 3 ft/sec. Table 8–5 compares corrosion rates at different velocities in 3.4% brine. This test was conducted on tubes. Table 8–6 shows similar data in natural seawater, and Table 8–7 includes constant immersion data in seawater.

C. GALVANIC CORROSION

When considering service in seawater, the engineer must concern himself with *galvanic corrosion*. The voltage and current flow that results from coupling dissimilar metals can lead to rapid and catastrophic attack of the more anodic component. Figure 8–4 shows a galvanic series of materials in seawater. In general, when joining dissimilar metals or alloys from this series, the engineer must select materials which lie close together in the series. Materials which are widely separated in the table should never be joined. Since coppers and copper alloys are cathodic to most engineering metals, galvanic corrosion is not likely to be a problem with them. The possibility of this type of corrosion must not be forgotten, however, particularly when the area of the anodic metal in the couple is much less than the cathodic area.

TABLE 8-4
Corrosion Rates in Brine
Instantaneous Rates in 3.4% Brine Solution at 120 Days. Liquid Cascading in
Laboratory Trough at About 3 ft/sec.

Alloy	Temper	Corrosion Rate, mils/y
C12200	Soft	1.0
C19400	1/2 Hard	0.08
C19400	Hard	0.24
C19500	Spring	<0.10
C23000	Soft	0.09
C26000	Soft	0.23
C42200	Hard	0.20
C44300	Soft	<0.10
C63800	Extra Hard	0.30
C66400	Extra Hard	0.11
C68700	Soft	<0.10
C68800	Soft	0.0002
C70600	Soft	0.07
C71500	Soft	0.06
C72500	Soft	<0.10
C76600	Soft	<0.10
C77000	Soft	<0.03

TABLE 8-5
Effect of Solution Velocity on Corrosion Rates in Brine Instantaneous Rates in 3.4%
Brine Circulating in Tubes. Temperature, 40°C (104°F); Time, 365 Days.

Alloy	Velocity, ft/sec	Corrosion Rate mils/yr
C12200	3.6	0.18
	5.0	0.40
	11.5	0.35
C19400	3.6	0.32
	5.0	0.10
	11.2	0.10
C19500	7.0	0.30
C44300	3.7	0.27
	5.4	0.45
	12.5	0.77
C68700	3.6	0.40
	5.2	0.10
	11.7	0.20
C70600	7.0	0.10

TABLE 8-6
Corrosion Rates in Cascading Seawater
Instantaneous Rates in Cascading Natural Seawater, Trough Test at Daytona Beach.

Alloy	Temper	Time, days	Corrosion Rate, mils/yr
C11000	Hard	156	0.92
C26000	Soft	—	0.48
C42200	Hard	150	1.00
C51000	Soft	365	0.22
C70600	Soft	365	0.04
C77000	Soft	365	0.90

TABLE 8-7
Corrosion Rates Submerged in Seawater. Instantaneous Rates at 365 Days. Total Immersion in Seawater at Daytona Beach.

Alloy	Corrosion Rate, mils/yr
C11000	0.30
C19400	0.16
C26000	0.12
C42200	0.30
C51000	0.43
C70600	0.02

D. BIOFOULING

Biofouling is most often a problem in marine environments. Marine organisms, such as barnacles and oysters, attach themselves to a metal surface and effectively seal off a small portion of the surface from the environment. The process is aggravated by warm water, which promotes rapid growth. Rapid corrosion can occur beneath the organism in materials which are sensitive to crevice corrosion. Copper provides excellent resistance to biofouling, and for this reason copper alloys have been the choice in marine service for many years. Small amounts of copper ions in solution at the seawater interface are very effective in preventing the attachment of marine life. Alloying additions which decrease the corrosion rate tend to lower the fouling resistance of copper alloys. However, all copper alloys have relatively good biofouling resistance compared to other materials. The copper–nickels are resistant to both fouling and pitting, and therefore are most useful in

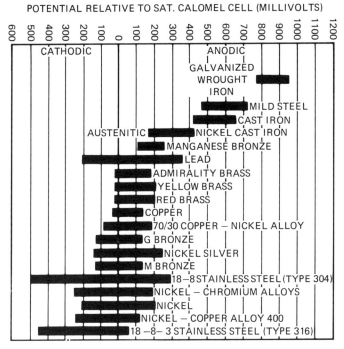

FIGURE 8–4
Galvanic series in seawater. (LaQue The International Nickel Company, Inc.)

seawater environments. Alloy C70600 (copper-nickel, 10%) has the best combination of biofouling and pitting resistance. Table 8–8 gives relative rankings of copper alloys and some other metals in regard to their resistance to biofouling.

E. EROSION-CORROSION

The relative velocity between a metal surface and a flowing environment can be a significant factor in determining the corrosion rate. In general, for copper alloys, the susceptibility to *erosion-corrosion* increases slowly with increasing flow velocity until a critical point is reached. This critical point is commonly referred to as the *breakaway velocity,* and is that velocity at which the protective corrosion product film is removed by erosion as fast as it can form. Syrett and Lapple [4] have described a mechanism of erosion-corrosion for copper alloys which produces a

TABLE 8-8
Relative Biofouling Resistance of Materials in Quiet Seawater

Alloy	Resistance to Biofouling
Copper	
Alloy C14200 (phosphorus deoxidized, arsenical)	Best
Alloy C19400 (phosphorus deoxidized, iron)	Best
Alloy C70600 (copper-nickel, 10%)	Best
Alloys C44300-44500 (admiralty brass)	Best
Alloy C68700 (aluminum brass, arsenical)	Good
Alloy C60800 (aluminum bronze, 5%)	Fair
Alloy C71500 (copper-nickel, 30%)	Fair
Aluminum	Least
Carbon steel	Least
Stainless steel	Least
Titanium	Least

curve like that shown in Figure 8–5a. Beyond the breakaway velocity, the rate of attack increases rapidly in region C because of the combined effect of erosion and galvanic action between the large area of metal oxide (cathode) and small areas of metal surface (anode) exposed by erosion—thus the name "erosion-corrosion" for this type of attack. The rate of attack diminishes in region D because a greater percentage of metal surface is exposed, reducing the galvanic effect. Alloys such as the copper-nickels and aluminum bronzes which form thin, adherent films are the best choices for reducing erosion-corrosion in seawater.

Flow in condenser and heat exchanger tubes is uniformly turbulent under normal operating conditions. Areas of severe damage can occur when local conditions create high turbulence and impingement that exceeds the material's resistance. Once an impingement pit develops, the turbulence associated with the roughened surface may be sufficient to cause the attack to continue. Erosion-corrosion is generally of interest in such applications as heat exchanger or condenser tubes, propellers, rotors, and ships' hulls.

F. PITTING

When anodic areas on a metal surface remain stationary, rather than shifting about as during uniform corrosion, *pitting* occurs. Pitting in copper is frequently characterized by the formation of a large, well-defined pit beneath a cap or crust of corrosion products. This cap

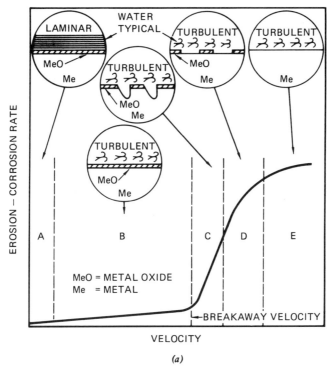

FIGURE 8–5a
Schematic of the changes in erosion-corrosion mechanism as velocity is increased. (Syrett and Lapple, (4))

prevents the corrosion products from being flushed from the pit. The acidity at the anode (pit) increases as more chloride migrates to the anode, and the corrosion rate increases with time. The corrosion reaction occurring in the pit thus becomes self accelerating unless the cap is removed and the pit flushed out. Sometimes this will arrest the attack.

As noted earlier, the coppers and copper alloys depend on the formation of a protective oxide film for corrosion resistance. When there is insufficient oxygen to maintain the film, it can break down and local pitting occurs. Oxygen deficiency can develop in narrow crevices where the liquid medium does not circulate and is cut off from air, or it may occur below surface deposits. Chemical concentration cells may form and cause local pitting due to electrochemical action. Corrosion pitting is particularly harmful because the attack is concentrated in a small area and can result in perforation and leaking of containers.

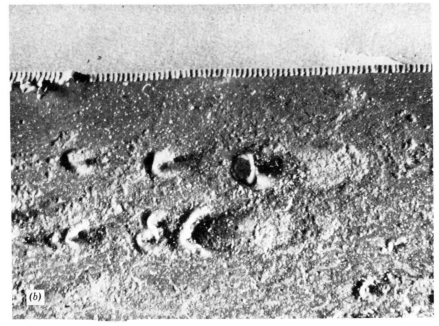

FIGURE 8–5b

Inlet erosion-corrosion on C19400 condenser tube due to turbulence in sea water. Note horseshoe shape of pits.

G. STRESS CORROSION

The combined effect of corrosion and static stress produces a form of attack in many different metals in which cracks develop at right angles to the direction of the applied stress. For this *stress-corrosion* attack to occur, it is necessary to have corrodent, tensile stresses at the surface and a susceptible alloy system.

This sort of corrosion failure in brass has been commonly called *season cracking.* The term comes from the outbreak of cracking of cartridge cases observed during the hot, rainy seasons in former British colonies. The cracks actually may be intergranular (following a path along the grain boundaries) or transgranular (crossing through the grains). See Figures 8–6a and b.

Certain homogeneous alloys, such as the copper–nickels, are highly resistant to stress corrosion. On the other hand, the high brasses, copper-zinc alloys, are the most susceptible. The potential for this type of corrosion must be considered when designing brass parts. Typical stress-corrosion-provoking environments are ammonia, nitrogen compounds, and chlorides. Ammonia-containing cleaning solutions and

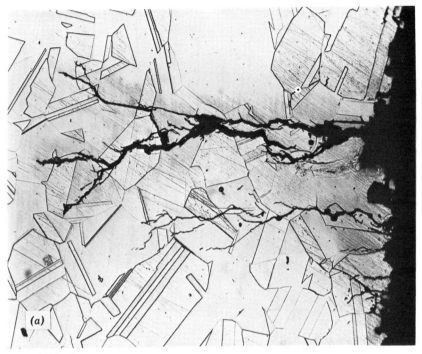

FIGURE 8–6a
Photomicrograph showing transgranular stress-corrosion cracking of Alloy C77000 after exposure in moist ammonia. Magnification: 250x.

plastic molding compounds which give off ammonia when heated can cause cracking. Storing metal in areas near process chemicals or in places where such vapors may be present should be avoided. Even something as unlikely to be suspected as rat urine can cause stress-corrosion cracking of brass.

The stresses which can lead to this type of cracking may arise from externally applied loads or from internal *locked-in, residual stresses* that develop during fabrication. Of the two sources, residual stresses are the most commonly associated with cracking. In general, applied service loads are usually anticipated during design and their extent and direction can be readily calculated and controlled. Residual stresses are often difficult to measure and do not signal their presence until cracking occurs. They develop as the result of a variety of factors, including cold deformation of any kind, such as drawing, forming, stamping, shearing, or bending. They also may result from nonuniform or local heating or from differential cooling. Sometimes, mechanical damage such as dents or scratches is responsible.

Nonuniform or unbalanced internal stresses which result from some

FIGURE 8–6b
Photomicrograph showing intergranular stress-corrosion cracking of Alloy C23000 after exposure to pH 7 Mattsson solution. Magnification: 250x.

types of forming, such as hollow sinking or reducing of tubes, can promote stress-corrosion susceptibility. A large die entry angle and small die bearing surface produce a high level of residual stress. Heavy reductions create smaller stress gradients than do medium reductions of about 20 to 30%. With a properly designed drawing operation it is possible to draw practically any size brass tube to any desired degree of hardness and still control the locked-in stresses enough to provide low stress-corrosion cracking susceptibility.

While the possibility of stress corrosion must always be a consideration when designing brass parts, a great deal is known about this type of corrosion and it is easy to control. For example, copper alloys containing less than 15% zinc (C23000, C22600, C22000, C42200, C66400) are far less susceptible than those containing a higher percentage of zinc (C24000, C26000, C26200, C26800, C76200, C77000). When service stresses are high, alloy selection is often the best means of control. Table 8–9 contains data showing the relative resistances to cracking in industrial and marine atmospheres of a number of alloys in various tempers.

Stress-relief annealing is a useful, inexpensive means of reducing stresses developed during fabrication. To be effective, stress relieving must be done on the completely formed part. It is accomplished by heating in the temperature range of 245 to 530°C(475 to 1000°F), depending on the alloy.

A less attractive and often impractical way to control stress corrosion is by the removal of the responsible corrodent. For example, ammonia- and chloride-containing lubricants can be used to draw brass parts, but only if they can be thoroughly removed immediately after drawing. Cleaning baths and other chemical processing equipment which give off corrosive fumes should not be located near stamping presses. Even sawdust on a damp concrete floor has been known to cause season cracking. Situations such as these have, on occasion, been the source of seemingly "mysterious" cracking problems with brass.

Plating, discussed in a later section, is often used as a stress-corrosion inhibitor on brasses. The most common kinds of plating used for this purpose are tin, hot-dipped or electroplated, and nickel. While generally effective, plating does not always insure resistance to cracking because of the presence of voids in the coating.

Research has been going on for many years and is currently continuing to establish test methods to determine stress-corrosion susceptibility and correlate them with field experience. The mercurous nitrate test described in ASTM B154 has long been in use to check copper alloy parts after fabrication to estimate if the stress level is high enough to cause susceptibility to failure. When it appears that parts may be susceptible to stress cracking, they can be relief annealed or plated, or an alloy which is not susceptible can be substituted. However, this test is not suitable to determine the susceptibility of alloys.

Stress-corrosion susceptibility tests are generally conducted with a standard horseshoe specimen, as shown in Figure 8–7. These specimens are exposed to atmospheric conditions, such as a rooftop in an industrial area, or to accelerated conditions in a laboratory, suspended in the vapors over moist ammonia. Table 8–9 shows such data for tests in the atmosphere. Table 8–10 provides laboratory test results comparing the performances of several copper alloys in three different pH ranges of Mattsson's solution and moist ammonia. Mattsson's solution is a laboratory test solution containing copper and ammonium sulfate in ammonium hydroxide. It shows pH sensitivity to stress-corrosion cracking in brasses and has been found to correlate well with service conditions. Additional information regarding the use of Mattsson's solution is available in ASTM G37.

TABLE 8-9
Atmospheric Stress-Corrosion-Cracking Data

Alloy	Temper	New Haven (Time to Fail)	Brooklyn (Time to Fail)	Daytona Beach (Time to Fail)
C11000	Soft	NF[a] in 7 yr	NF in 7 yr	NF in 6 yr
C11000	Hard	NF in 3 yr	NF in 3 yr	NF in 3 yr
C19400	Extra Hard	NF in 2 yr		
C19500	Spring	NF in 4 yr	NF[b] in 4 yr	NF in 4 yr
C23000	Hard	NF in 7 yr	NF in 7 yr	NF in 7 yr
C26000	Hard (T)[c]	37 days	12 days	NF in 3 yr
C26000	Hard (L)[d]	1 of 3 in 113 days		
C42200	Hard	NF in 7 yr	NF in 7 yr	NF in 7 yr
C42500	Extra Hard	NF in 2 yr	NF in 6 mo	NF in 6 mo
C51000	Extra Hard	NF in 7 yr	NF in 7 yr	NF in 5 yr
C52100	Hard	NF in 9 yr	NF in 9 yr	NF in 9 yr
C63800	Extra Hard	NF in 8 yr	NF in 6.5 yr	NF in 7 yr
C66400	Extra Hard	NF in 3 yr	NF in 2 yr	NF in 3 yr
C68700	Hard	401 days avg.	3 of 5 in 1000 days	
C68800	Hard	4 of 5 in 1500 days	2 of 5 in 1500 days	NF in 10 yr
C68800	Hard (HT)[e]	1325 days avg.		
C70600	Hard	NF in 4 yr	NF in 3 yr	NF in 4 yr
C72500	Hard	NF in 4 yr	NF in 3 yr	NF in 4 yr
C75200	Extra Hard	NF in 3 yr	NF in 3 yr	NF in 3 yr
C76200	Extra Hard	196 days avg.	266 days avg.	NF in 2 yr
C76600	Extra Hard	266 days avg.	132 days avg.	1 of 5 at 754 days
C77000	Extra Hard	277 days avg.	515 days avg.	4 of 5 at 904 days

[a]NF = no failures in exposure time given.
[b]Blank = not tested.
[c]T = transverse.
[d]L = longitudinal.
[e]HT = aged.

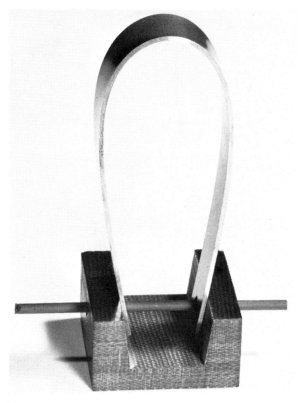

FIGURE 8–7

Typical horseshoe type stress-corrosion test specimen. Preformed U-bend specimen with legs compressed in Micarta jig to stress outer bend fibers in tension.

TABLE 8-10
Stress-Corrosion Resistance of Copper Alloys
Time to Failure (Cracking) in Hours.

Alloy	Temper	Mattsson pH 4	Mattsson pH 7.2	Mattsson pH 10.0	Moist NH₃
C11000	Hard	—	NF	—	NF
C19400	Extra Hard	NF[a]	NF	NF	NF
C19500	90% CR[b]	NF	NF	NF	NF
C23000	50% CR	NF	39.0	97.0	3.2
C26000	Hard	398	5.5		1.5
C26000	Extra Hard		6.3	136	
C42200	Extra Hard	NF[c]	1006		14.0
C42500	Extra Hard	NF	1143		50.9
C44300	40% CR	83	29	38	1.3
C51000	50% CR		NF		273
C63800	Extra Hard	NF	NF	617	30.5
C66400	40% CR		NF		17.0
C68700	40% CR	415	NF	123	0.9
C68800	40% CR	NF	640	134	1.0
C70600	50% CR		NF		NF
C72500	40% CR		NF		NF
C75200	Extra Hard	887	884		
C76200	50% CR	250	60	130	90
C77000	50% CR	252	250	239	309

[a]NF = no failures.
[b]CR = cold reduction.
[c]Blank = not tested.

H. CORROSION FATIGUE

The fatigue strengths of most copper alloys are well known. This term is defined as the stress which can be endured by the metal for a given number of cycles in air before it cracks by fatigue. Under corrosive conditions, the same metal will fail at much lower stresses. This combined action of corrosion and repeated stress is known as *corrosion fatigue*. The corrosion fatigue properties of a metal depend on its corrosion resistance in the particular environment in which it is located. Even though heat treatment and cold work may improve the mechanical properties and fatigue strength in air, the resistance to corrosion fatigue is improved only when the corrosive conditions are reduced. Consequently, even though a metal has a higher fatigue strength when it is cold worked, it is still susceptible to corrosion fatigue. Alloying is of benefit only when the corrosion resistance of the alloy is increased.

As stated previously, the corrosion resistance of copper alloys is improved by the formation of an oxide film on the metal surface. These films are generally brittle and have low strength. When they are ruptured by the action of cyclic stresses, corrosion proceeds at an increased rate at the breaks in the film and frequently forms sharp pits which act as stress risers of unusually high intensity. It is at the base of these pits that cracks are initiated and usually progress in a transgranular mode until failure occurs. When selecting materials to resist corrosion fatigue, the corrosion resistance of the metal in the environment under consideration is most important.

I. DEZINCIFICATION

The corrosion of brass by selective removal of zinc is a common type of attack, particularly in seawater and soft waters containing carbon dioxide. The surface area of the original alloy which is attacked is converted into a spongy mass of copper which has poor mechanical strength. This type of attack is called *dezincification* and is accelerated by poor aeration, such as would be found in crevices (screw fittings) and beneath deposits of debris or corrosion products. The attack is also accelerated by high temperature and either acid or very strong alkaline conditions. The only outward sign of attack is a change in color and texture. Instead of the yellow color typical of brass, the affected areas have the red color of copper and are spongy and soft. Two theories on the mechanism exist: either that the zinc is dissolved preferentially, leaving the copper behind; or (more favored) that both metals dissolve and the copper is subsequently redeposited. The zinc-rich brasses are the most susceptible to this form of attack. Alpha brasses with a zinc content of 15% or less, such as Alloy C23000, are very resistant. See Figure 8–8.

There is no reliable method of completely inhibiting dezincification in the high zinc brasses, but small additions of arsenic (0.02 to 0.06%), antimony, or phosphorus are successful in preventing the dezincification of alpha brasses in most types of water. Of the three inhibitors, arsenic is the most widely used and is the only inhibitor added to the aluminum brasses, which are also susceptible to dezincification.

J. INTERGRANULAR CORROSION

This type of attack, sometimes referred to as *intercrystalline corrosion,* is associated with a difference in the corrosion potentials of the grain boundary or the grain-boundary region and the rest of the grain. The grain-boundary region is anodic, and the attack can be very severe owing

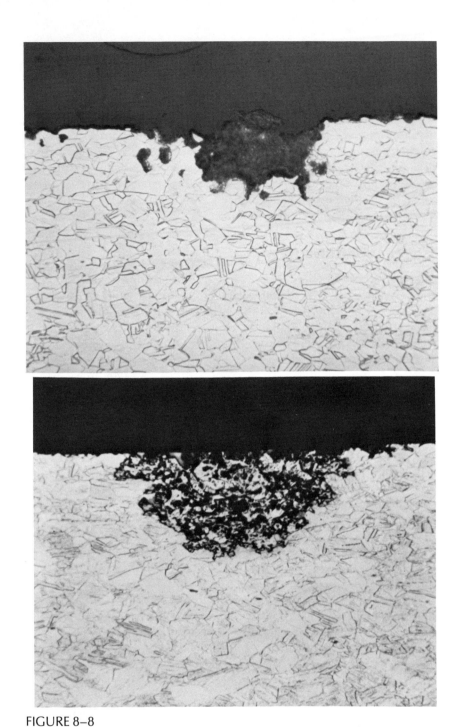

FIGURE 8–8
Photomicrograph of plug type dezincification in Alloy C26000 after 4 years' exposure in Brooklyn, New York. Magnification: 250x.

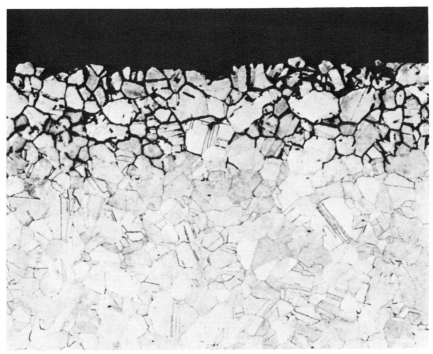

FIGURE 8-9

Photomicrograph of intergranular corrosion in Alloy C44300 condenser tube due to contaminated sea water. Magnification: 250x.

to the large cathode (grain) to anode (grain-boundary) area. In a given alloy the smaller the grain size, the lower is the susceptibility to attack. As noted previously, when cracking occurs owing to the combined action of corrosion and mechanical stress, it is usually classified as stress-corrosion cracking. In some cases, the precipitation of minor alloying constituents as intermetallic compounds at grain boundaries can result in the grain boundaries becoming more anodic than the surrounding grains. An example of this is the precipitation of iron–nickel-rich particles in the grain boundaries of C70600 because of improper annealing temperature, which leads to corrosion in environments such as seawater and high-temperature steam. See also Figure 8-9.

BIBLIOGRAPHY

1. *Metal Finishing*, 41st ed., 1973, pp. 210–211.

2. North, R. F., and M. J. Pryor, *Corros. Sci*, **10**, 297–311, (1970).

3. Pilling, N. B., and R. E. Bedworth, *J. Inst. Met.*, **29**, 534, (1923).

4. Stanford Research Institute, *Erosion-Corrosion of Copper-Nickel Alloys*, 1974.

9

ELECTRICAL CONTACT RESISTANCE

Copper alloys are widely used in applications in which electrical contact is made between two members. A broad spectrum of contact types exists, ranging from those in which precious metal inserts (gold or silver) are used, those in which a precious-metal-plated surface is used, and those employing a tin-plated surface, to those in which the base alloy surface is used as the contact surface. This chapter reviews significant factors influencing this important property and provides specific data on contact resistance. The use of platings and methods of applying them are also discussed.

STATIC CONTACT RESISTANCE

Contact resistance usually increases with lengthening exposure to the surrounding environment as a reflection of the gradual buildup of corrosion products on the surface. The extent of time-dependent increase in contact resistance varies from alloy to alloy.

Contact resistance varies significantly with contact pressure. As might be expected, at low contact pressures the differences between alloys and the effect of degradation due to aging are more pronounced.

Olin's Metals Research Laboratories have performed *static contact resistance* measurements upon many coppers and representative copper alloys. These measurements were made on bare metal in the freshly cleaned condition and after storage for time intervals up to 5000 hours. They were also made at several contact pressures.

The probe used in these tests was a stainless steel rod with a hemispherical tip of 0.2 in. (0.5 cm) diameter. The probe and its attendant electrical connections were mounted in the sample pan of a triple-beam balance. The sample was held rigidly above it, and when the balance was released, the probe contacted the sample. The balance was zeroed so that, by moving its counterweights, the probe could be made to contact the specimens with any desired load in the range of 10 to 1000 gs. A dashpot was attached to the balance to damp out stray vibrations, which become intolerable at low (10 to 50 g) loads; it also served to control the rate of loading.

A 12 V storage battery provided a constant voltage source for the resistance measuring circuit. The circuit was arranged to provide an open circuit potential of 1 V across the specimen. The current through the specimen was approximately 7 M.A. Resistance was determined by simply measuring the actual current through and voltage drop across the specimen. Five readings at each load, each at a different point on a sample, were averaged to provide a contact resistance value.

Mill produced strip of normal characteristics was used in all tests. Test specimens were cleaned just before the test started, using solutions normally used in mill cleaning. Similar specimens were stored in drawers to avoid excessive accumulation of dust but not otherwise protected. These were withdrawn from storage and, without cleaning or other treatment, tested after 1000 hours and 5000 hours in storage. Table 9–1 provides contact resistance data for several copper alloys for loads ranging from 10 to 1000 gs.

DYNAMIC CONTACT RESISTANCE

While static contact resistance is useful for making judgments regarding the relative performances of materials, some designers claim that no junction is truly static. The slight movements induced in the junction by vibrations, movements, and temperature changes, in effect, make the junctions dynamic. A test similar to the static contact resistance test, except that it provides for movement of the contact probe, is used to determine the *dynamic contact resistance*. This movement results in a more rapid formation of oxide at the probe and a resultant increase in contact resistance. Data are being collected and attempts are being made to correlate this test with actual field results.

PLATING

The requirements for efficient electrical contacts are that they successfully close the circuit, carry the current without overheating, and interrupt the current without deterioration. In some types of electrical

TABLE 9-1
Static Electrical Contact Resistances (Ω) of Various Alloys after Aging and at Different Loads

Alloy	Age (hours)	Load (g)					
		10	20	50	100	200	1000
C11000	0	0.017	0.013	0.009	0.009	0.007	0.006
	1000	0.019	0.014	0.011	0.011	0.008	0.006
	5000	0.023	0.014	0.010	0.010	0.008	0.006
C19400	0	0.018	0.011	0.007	0.006	0.005	0.006
	1000	0.021	0.017	0.011	0.010	0.008	0.007
	5000	0.023	0.015	0.011	0.010	0.010	0.005
C19500	0	0.011	0.009	0.007	0.006	0.005	0.004
	1000	0.027	0.012	0.010	0.009	0.008	0.006
	5000	0.047	0.018	0.016	0.009	0.008	0.006
C26000	0	0.022	0.013	0.010	0.008	0.007	0.006
	1000	0.022	0.015	0.012	0.010	0.008	0.006
	5000	0.025	0.015	0.011	0.009	0.008	0.005
C42200	0	0.026	0.019	0.012	0.009	0.007	0.004
	1000	0.031	0.022	0.016	0.012	0.010	0.005
	5000	0.056	0.024	0.013	0.011	0.008	0.005
C42500	0	0.028	0.020	0.014	0.012	0.009	0.006
	1000	0.034	0.023	0.017	0.013	0.011	0.007
	5000	0.047	0.023	0.016	0.013	0.011	0.006
C51000	0	0.019	0.015	0.011	0.008	0.006	0.003
	1000	0.022	0.017	0.013	0.008	0.007	0.003
	5000	0.030	0.017	0.012	0.008	0.007	0.003
C66400	0	0.022	0.015	0.009	0.006	0.005	0.003
	1000	0.026	0.019	0.013	0.008	0.007	0.004
	5000	0.045	0.019	0.009	0.009	0.008	0.004
C63800	0	0.020	0.018	0.012	0.010	0.008	0.004
	1000	0.022	0.019	0.014	0.012	0.008	0.004
	5000	0.021	0.018	0.012	0.009	0.007	0.004
C68800	0	0.017	0.013	0.009	0.008	0.006	0.004
	1000	0.019	0.011	0.009	0.007	0.006	0.004
	5000	0.017	0.012	0.009	0.008	0.007	0.004
C72500	0	0.018	0.017	0.011	0.019	0.017	0.003
	1000	0.020	0.016	0.013	0.019	0.016	0.003
	5000	0.047	0.018	0.013	0.019	0.019	0.003

or electronic equipment, the contact is required to meet these requirements through many thousands of make and break cycles during its lifetime. Consequently, the material must resist tarnishing and oxidation to preserve low contact resistance. The temperature rise of the junction will depend on this resistance and on the ability of the materials to dissipate heat. All of these requirements have led to plating contact with precious, semiprecious, or highly corrosion resistant metals.

The need for very stable, low resistance junctions has grown with the use of solid-state circuitry, which often employs devices having low contact forces and operating at low voltages and amperages. From this basic requirement, the term *dry circuit* has evolved. This term is used to denote conditions in which the voltage is too low to puncture insulating films in the contact pair. Dry circuit limits are generally 1 V or less and 1 A or less. Contact force, corrosion protection, shelf life, solderability, oxidation resistance, environment, and application all play an important part in the requirements for plating on electrical contacts.

PLATING MATERIALS

Gold has been the plating material most used for high reliability contacts, although silver is gaining in use because of lower cost. With gold, there is no reaction with other metals, or with the atmosphere to form oxide and tarnish. Because of these properties, it is the preferred plating material for dry circuit contacts. It will work with very low contact forces since there is no need to break through oxide films. Because of low contact forces, many engagements and disengagements can be made without wearing through the plating. Pure gold is relatively soft, and in the soft unalloyed state it has its best electrical conductivity, corrosion resistance, and soldering properties. Its wear resistance is very poor, and it will not survive many cycles of operation if the contact forces are large. Gold plating on copper alloys must be thick enough to be free of pores, and it must be free from impurities so the thickness can be kept at a minimum. To help meet these requirements, a *nickel diffusion barrier* is frequently plated on prior to the gold plating. Such a barrier functions, in part, to prevent copper atoms from gradually diffusing into the very thin gold film, particularly at elevated temperatures, and affecting its properties.

When the requirements are not so stringent, or when costs are important, tin and tin-lead alloys (solder coating) are considered. In fact, these coatings are becoming increasingly popular in applications where it is now recognized that gold is more than what is actually needed. Tin and tin-lead are relatively inexpensive, and even though they do form oxides in air, they can still be used because of the relative ease with which this film can be mechanically broken. This allows low resistance metal-to-metal contact, but certain minimum forces are required to break this film. These platings have good solder and wetting characteristics and are relatively immune to many pollutants.

Recent test data indicate that in some applications bright tin-lead deposits perform as well as gold. Such bright platings are achieved by the addition of special agents to the plating baths, and it appears that this can improve their performance. Tin and tin-lead coatings do have certain limitations, and they are not recommended for long-duration, repeated mate and unmate conditions. They are rarely used for dry circuit contacts. If the engagement force is important, tin and tin-lead alloys may suffer because they generally have higher engagement forces than gold. Nicotera [2] has done considerable rating of tin, tin-lead, nickel, gold, and combinations of these platings. Table 9–2 provides ratings for these platings after humidity exposure, sulfur environment exposure, and salt fog exposure. Table 9–3 shows the influence of wear testing. Table 9–4 shows how some of the coatings affect engagement and withdrawal force.

Nickel has gained its widest use as a plating on copper alloys principally as a diffusion barrier, as noted previously. Consideration must be given to the properties of this nickel diffusion barrier if the contact must meet certain forming or bend cycle requirements. Under certain conditions, this nickel diffusion barrier can be brittle and result in an early fracture during forming or bend cycle testing, and this fracture in the nickel diffusion barrier acts as a stress riser, causing the contact lead to fail prematurely. Reduced performance of copper alloys in electrical lead bend test has been traced to excessively thick nickel barrier plating.

Other elements, such as palladium, rhodium, and platinum, have been used occasionally as plating materials, but their application is very specialized and so restricted that it will not be covered here.

METHODS OF PLATING

METAL DIPPING. Metals may be plated by *dip coating* in a molten bath of another metal. This method, the oldest method of plating, is usually restricted to coating with metals of low melting point, such as tin, tin-lead, or lead. The surface of the metal to be *hot dipped* must be cleaned to insure adequate adhesion of the coating metal. The control of impurities in the molten bath is also very important.

TABLE 9-2
Effects of Exposure of Plated Contacts to Various Environments on Their Contact Resistance

Plating on Connector	Plating on Mating Pin	Initial No Exposure	Rating[a] after Exposure to Humidity[b]	Sulfur[c]	Salt Fog[d]
Tin/lead	Tin/lead	A-1	A-1	A-1	A-6
Tin over copper	Tin	A-2	—	A-4	A-1
Tin/lead, reflowed	Tin/lead, reflowed	A-3	A-4	A-2	R-22
Bright tin/lead	Bright tin/lead	A-4	—	A-8	A-3
Bright tin	Bright tin	A-5	R-17	A-3	R-18
Tin over copper, reflowed	Tin	A-6	—	A-6	R-12
Tin, reflowed	Tin	A-7	—	—	R-9
Tin over copper	Gold	A-8	A-7	A-5	M-7
Tin/lead, reflowed	Gold	A-9	M-11	A-17	R-15
Tin/lead	Gold	A-10	A-8	A-11	R-20
Tin over copper, reflowed	Gold	A-11	A-2	A-9	M-8
Bright tin/lead	Gold	A-12	—	A-15	R-10
Duplex gold	Gold	A-13	—	A-14	A-2
18K gold	Gold	A-14	—	A-16	A-4
Bright tin	Gold	A-15	A-5	A-13	R-11
Gold	Gold	A-16	A-6	A-12	A-5
Tin, reflowed	Gold	A-17	A-3	A-10	R-13
Nickel/boron	Nickel/boron	A-18	M-10	M-19	R-16
Nickel	Gold	A-19	—	M-18	R-17
Nickel/boron	Gold	M-20	M-9	M-20	R-19
Tin/nickel	Tin/nickel	M-21	R-18	R-23	R-23
Tin/nickel	Gold	M-22	R-12	R-21	R-21
Nickel	Nickel	M-23	—	R-22	R-14
Tin reflowed	Tin reflowed	—	R-14	—	—
Tin over copper, reflowed	Tin over copper, reflowed	—	R-16	—	—
Tin, reflowed	Tin	—	—	A-7	—

[a]Rating: A = less than 5 mΩ; M = greater than 5 but less than 10 mΩ; R = greater than 10 mΩ. Number after rating letter shows standing of pair within each column.

[b]Fifty 24 hour cycles in controlled humidity chamber: 1 cycle = 4 hours 50% max. R.H., 25 ± 5°C; 4 hours 90–90% R.H., 75 ± 5°C; 16 hours 50% max. R.H., 75 ± 5°C.

[c]Exposed in desiccator containing flowers-of-sulfur over water for 10 days at 80 ± 3% R.H. and 65 ± 2°C.

[d]Exposed to 20% salt fog for 100 hours per QQ-M-151A, (automatically controlled chamber).

TABLE 9-3
Effect of Wear Testing[b] on Contact Resistance[c]
(Listed in Order of Increasing Resistance)

Rating[a]	500 Cycles	Rating	1000 Cycles
A	Bright tin	A	Tin, reflowed
A	Tin over copper, reflowed	A	C.I. gold
A	C.I. gold	A	Bright tin
A	Tin, reflowed	A	Tin/lead
A	Tin/lead	A	Nickel-boron
A	Tin/lead, reflowed	A	Tin over copper
A	Nickel/boron	M	Tin over copper, reflowed
A	Tin over copper	M	Tin/lead, reflowed
R	Tin/nickel	M	Tin/nickel

[a]Rating: A = less than 5 mΩ resistance; M = greater than 5 but less than 10 mΩ resistance; R = greater than 10 mΩ resistance.
[b]One engagement and disengagement of a 0.025 in. sq. steel pin of 4 μin. finish is one wear cycle. Cycling performed in automatic cycling fixture.
[c]All samples are wear tested with steel pin and electrically tested with gold-plated pin.

TABLE 9-4
Engagement-Disengagement Force

Connector Plating	Engagement Force Initial—0.026 in. sq. steel pin (average value, g)	Disengagement Force— 0.024 in. sq. steel pin after conditioning cycles with 0.026 in. sq. pin (average value, g)
Tin/nickel	162	57
Tin over copper	176	57
Tin/lead	155	61
Tin/lead, reflowed	204	61
Gold	144	61
Tin, reflowed	183	62
Bright tin	188	63
Nickel/boron	172	66
Tin over copper, reflowed	210	68

Some advantages of hot-dip plating are that the coating is soft, formable, and free from porosity.

ELECTROPLATING. Electroplating is the deposition of an adherent metal coating upon an electrode from a plating solution by an electric current. Securing a surface with properties or dimensions different from those of the base metal is the usual purpose of *electroplating.* It has an advantage over metal dipping in that a wide variety of metals can be electrodeposited on many different metal surfaces. For the engineer, it is extremely flexible, and either original strip or finished parts can be plated. It also provides the capability to selectively plate certain areas, such as contact surfaces, of the finished parts. Another advantage is that a uniform thickness can be deposited, and there is greater control over the thickness than if it is deposited by metal dipping. This is important when a precious metal such as gold or silver is being plated. Electroplating also gives the engineer some flexibility in the surface finish of the coating. Typical metals which are plated on copper alloy contact members are tin, tin-lead (solder), nickel, and gold over a nickel barrier plate.

During electroplating, ionic hydrogen is present near the metal surface. Under certain circumstances, this form of hydrogen can combine with oxygen in some alloys and lead to a problem known as hydrogen embrittlement. Care must be taken to insure that the plating bath is compatible with the material to be plated in order to avoid such embrittlement.

The preparation of copper alloys for electroplating is an important consideration. In order for electroplating to be successful the metal surface must be chemically clean and free from oxide and scale. This may involve metal surface preparation, such as polishing, brushing, buffing, electropolishing, solvent cleaning, and other preliminary operations. These pretreatments prior to plating are typically divided into two stages, preliminary and final treatment. The bulk of cleaning is done during the preliminary phases, which involve the major removal of oil, grease, buffing or drawing compounds, and the removal of scale, heavy oxide, and other heavy soils. The final treatment usually removes the last traces of any oil and grease, and conditions the surface for the electroplating bath. Any acid dips used in the final stage should not be expected to remove heavy scale or soil. Their basic and only purposes are to neutralize the last traces of an alkaline cleaner remaining after rinsing and to activate the surface for plating. Care should be taken to select acids which do not readily form compounds with alloying ingredients in the metal to be plated. Table 9–5 gives typical pretreatments for use on copper and copper base alloys.

TABLE 9-5
Typical Pretreatment Procedures for Coppers and Copper Alloys[a]

Cycle V-1	Cycle V-2	Cycle V-3 (for soldered parts or leaded brass)	Cycle V-4	Cycle V-5
(1) Preliminary pretreatment	(1) Preliminary pretreatment	(1) Preliminary pretreatment	(1) Preliminary pretreatment	(1) Cathodic or soak clean
(2) Rinse	(2) Rinse	(2) Rinse	(2) Rinse	(2) Rinse
(3) Electroclean[b]	(3) Electroclean[b]	(3) Cathodic clean	(3) Electroclean	(3) Acid dip—HCI 50%
(4) Rinse	(4) Rinse	(4) Rinse	(4) Rinse	(4) Rinse
(5) Acid dip[c]	(5) Sodium cyanide dip, 2–6 oz/gal	(5) Acid dip, 10–20% fluoboric acid	(5) Bright dip	(5) Plate
(6) Rinse	(6) Rinse, or copper strike and rinse	(6) Rinse	(6) Rinse	
(7) Plate	(7) Plate in alkaline solution	(7) Cyanide dip	(7) Rinse	
		(8) Rinse, optional	(8) Plate	
		(9) Cyanide copper strike		
		(10) Rinse		
		(11) Acid dip		
		(12) Rinse		
		(13) Plate		

[a] *For Nickel Plating.* Use Cycle V-1, V-3, or V-4. *For Copper Plating.* Use Cycle V-1, V-2, V-3, or V-4. *For Cadmium, Zinc, and Tin Plating.* Use Cycle V-1, V-2, V-3, or V-4. *For Silver Plating.* Use Cycle V-1, V-2, V-3, or V-4. *For Gold Plating.* Use Cycle V-1, V-2, V-3, or V-4. *For Lead Plating.* Use Cycle V-1, V-2, V-3, or V-4. *For Chromium Plating (Directly on Copper or Alloy).* Use Cycle V-1. For hard chromium use Cycle V-5.

[b] In many cases, proprietary cleaners are used in accordance with the supplier's recommendations. Anodic cleaning of copper and alloys has in recent years become preferred. Occasionally, cathodic followed by anodic cleaning in a separate tank is used. If the parts are tarnished and cathodic cleaning is to be used, it is advisable to remove the tarnish by an acid or cyanide dip prior to cathodic cleaning.

[c] The acid dip for copper or brass may be 5 to 10% by volume sulfuric acid, 10 to 20% by volume hydrochloric acid, or 10 to 20% by volume fluoboric acid, the last usually used when a fluoborate plating solution follows.

Because of the interaction of the various steps in plating on leaded brass, it is difficult to prescribe a cycle which will always give adherent electrodeposits. Cathodic cleaning is usually preferred, but very mild anodic cleaning is sometimes used. In many cases, fluoboric acid functions no better than sulfuric acid. Occasionally it has been found necessary to add carbonate to a new copper cyanide strike solution.

Data source see Reference 1.

ELECTROLESS CHEMICAL PLATING. The deposition of a metallic coating by a controlled chemical reaction which is catalyzed by the metal or alloy being deposited is typically referred to as *electroless* or *chemical plating.* One of the main advantages of chemical plating is that the deposits generally have lower porosity than electrodeposits; and, consequently, a thinner plating may do the same job as a heavier electrodeposition. The surface finish for parts to be chemically plated is extremely important, because the electroless plating operation does not have any leveling characteristics and so does not fill in any scratches or discontinuities in the surface. As a result, the plated surface is not smoother than the preplated surface. The microstructure of electroless plating is different and in some applications may be superior to that of electroplating. For example, the grain structure of chemically plated nickel is lamellar in structure, while the grain structure of electrodeposited nickel is columnar. A lamellar grain is a much more effective barrier to diffusion than a columnar grain. Chemical baths are autocatalytic and require no external electrical current. This eliminates the problem of high and low current density areas; thus intricate and irregularly shaped parts can be coated uniformly on all surfaces. Electroless nickel, plated on copper-alloys, has gained many electronic applications, such as printed circuit board patterns, connector contacts, semiconductor devices, and ceramic-to-metal integrated circuit packages. Electroless nickel may permit as much as 50% thinner gold plating, so its use may actually be more economical than that of electrolytic nickel on an overal basis.

BIBLIOGRAPHY

1. Graham, A. K., *Electroplating Engineering Handbook,* 3rd ed., Van Nostrand Reinhold Co., New York, 1971.
2. Nicotera, E. T., *Electronic Packaging and Production,* December, 1974.

10

FABRICATION BY FORMING, DRAWING, AND RELATED OPERATIONS

A large portion of the copper and copper alloy sheet and strip produced in brass mills is fabricated into useful parts by forming operations. Some parts, such as roll-formed brass tubes for automobile radiators and roll-formed and welded copper alloy tube for various heat exchanger and plumbing applications, use sizable amounts of strip, but the largest proportion undoubtedly is fabricated into useful parts by *press forming*. Included in this general term are such operations as *blanking, piercing, shaving, bending, forming, stamping, embossing, cupping, drawing, deep drawing, coining,* and the many variations of these, and the auxiliary press operations which supplement them.

There are many well-written texts on the subject of presses and the pressworking of metals, some of which are listed as references at the end of this chapter. The reader who is especially interested in the mechanical features of presses is referred to these. This chapter will be devoted to the characteristics of coppers and copper alloys that influence their pressworking and some of the defects and difficulties that cause problems. The four major features of strip that determine how the metal will respond in pressworking are (1) alloy or composition, (2) temper, (3) gauge or thickness, and (4) surface.

COMPOSITIONAL INFLUENCES

Chapter 4 described the basic metallurgy of the copper metals and the presence of different crystalline structures in different alloys. The basic crystal arrangement of copper and the single-phase copper alloy atoms is face-centered cubic. It was noted that this structure is very desirable, because it allows considerable plastic deformation without fracture. Indeed, the copper alloys are noted for their excellent pressworking characteristics. The combinations of strength, ductility, and work hardening rates of coppers and the common copper alloys such as cartridge brass, Alloy C26000, make them among the most workable metals from which parts can be fabricated in presses.

An example of the influence of composition on properties and formability is the effect of varying zinc contents in brasses. Copper C11000 strip (ETP copper—99.90% Cu), annealed for presswork, is a relatively low strength metal as compared to cartridge brass, Alloy C26000 (70% Cu—30% Zn). Its tensile strength is approximately 33 ksi, yield strength (0.2% offset) is approximately 6 ksi, and elongation in 2 in. is about 35%, depending on thickness. In contrast, the tensile properties of cartridge brass, Alloy C26000, with a comparable grain size of 0.030 mm would be tensile strength 50 ksi, yield strength 18 ksi, and elongation 56%. The properties of the copper-zinc alloys whose compositions lie between those of copper and cartridge brass, such as Alloy C21000 (5% Zn), Alloy C22000 (10% Zn), Alloy C23000 (15% Zn), and Alloy C24000 (20% Zn) with the same nominal grain size, lie between those of copper C11000 and brass C26000 (30% Zn). Figures 10–1 and 10–2 illustrate the differences in strength of these alloys. As would be

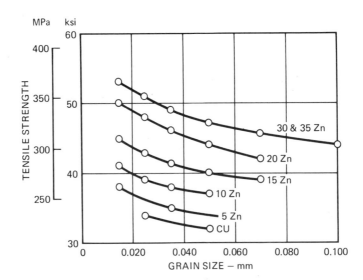

FIGURE 10–1
Effect of grain size on tensile strength of annealed 0.040 in. strips of copper and brasses of designated zinc contents. (Butts [4]).

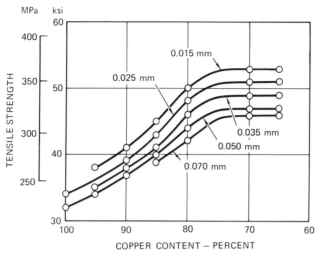

FIGURE 10–2
Tensile strength of 0.040 in. strips of copper and copper-zinc alloys of different copper contents annealed to designated grain sizes. (Butts [4]).

expected, the press drawing operations for the same part made from each of these might require some adjustment to allow for these differences in properties.

In draw press work on copper C11000, its lower strength and lower ductility in tension mean that its stretch formability is more limited than that of the brasses with higher elongation. In press-drawing copper, care must be exercised in designing and adjusting tools to avoid high localized stresses, because the low strength can be exceeded easily and the part being formed will fracture. Excessive pressure-pad hold-down must be avoided, and die radii must not be so small that they restrict metal flow into the die. Punch bottoms should be flat, and punch radii should be chosen to avoid both local stress and too much flow and stretch. In redraw operations, sidewall ironing should be used to keep stress distributed in the shell walls rather than concentrated on the bottom. These general statements apply to draw press operations on all relatively low strength coppers and the low zinc brasses.

Copper-zinc alloys that are high in copper, are all excellent for embossing and coining operations and for successive redraw operations, because their low yield strengths and low work-hardening rates allow them to readily fill the embossing die and take a sharp impression or to be deep drawn in a series of redraw steps. The United States penny is made from Alloy C21000 and is a good example of a low zinc copper alloy embossed part. Coins, medallions, jewelry, and hollow ware are among the many articles produced from high copper alloys of copper-zinc and coppers because of their malleability as well as their attractive colors.

The higher zinc brasses, such as Alloy C24000, low brass, Alloy C26000, cartridge brass, and Alloy C26200, high brass, having strengths comparable to those of the low carbon mild steels. Their strength is coupled with high ductility and a relatively low work-hardening rate. They are outstanding materials for deep drawing and other press forming operations, including stretch forming. With appropriate tool design and fabricating process design, there seems to be almost no limit to the formability of these alloys. The basic rule is to avoid applying excessive localized stress. When this is observed, the most intricately shaped drawn parts can be fabricated economically from these alloys.

Many other families of copper alloys also have good pressworking properties. Phosphor bronze, Alloy C51000 (Cu-5% Sn), has an excellent combination of high strength and high ductility. Thin walled shells are deep drawn using this alloy, annealed, and then corrugated to produce bellows of high fatigue strength and corrosion resistance and excellent flexibility. The nickel silvers (nickel brasses containing copper-nickel-zinc) are white copper alloys which, like the high zinc brasses, have excellent deep-drawing characteristics. However, they have somewhat higher work-hardening rates and require more interstage annealing than cartridge brass under similar circumstances. Some of the dispersion-strengthened alloys such as Alloy C63800 (Cu-3% Al-2% Si-0.4%Co) also have good deep drawability, similar to that of the nickel silvers, when in the fully annealed, soft condition. The high zinc leaded brasses when annealed are suitable for shallow-drawn parts, but the lead, depending on how well it is distributed in the structure, tends to have an embrittling effect. Garden-hose coupling nuts are a good example of the kind of part which can be press drawn from leaded brasses.

The copper-zinc-tin alloys such as C40500 (Zn 5%, Sn 1%), C41100 (Zn 9.5%, Sn 0.5%), C42200 (Zn 12%, Sn 1%), and C425000 (Zn 9.5%, Sn 2%) respond well to pressworking. Alloys C40500 and C41100 are similar to the high copper brasses, and C42200 and C42500 are similar to C24000 in deep-drawing properties. A special dispersion-strengthened alloy designed to have properties similar to these is C66400 (Zn 11.5%, Fe 1.5%, Co 0.5%, Cu bal.) There are many other special purpose coppers and copper alloys. By examining their compositions and mechanical properties carefully, and comparing them with standard alloys, the user can make a rough estimate of how they will respond in pressworking.

Figures 10–3 and 10–4, taken from the ASM *Metals Handbook* with added data for Alloys C19400, C63800, and C66400, show how work hardening decreases the ductility and increases the strength of copper and some copper alloys. Alloys with high elongation can be stretch formed. When they also have moderate work-hardening rates, they can

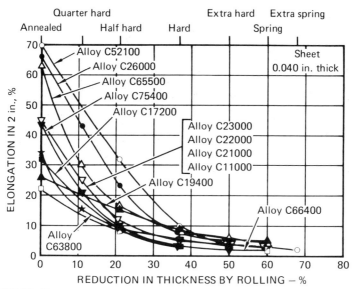

FIGURE 10–3

The decrease in elongation that occurs during cold rolling is shown for copper and copper alloys. Comparable decreases occur with equivalent percentages of area reduction in press drawing operations. (Reference 1).

be readily deep drawn. Alloys with low elongation and low work-hardening rates, such as C19400, C23000, and C22000, can also be deep drawn. Alloys with both low elongation and high work-hardening rates are more difficult to deep-draw.

In draw press operations and other forming methods for producing deep hollow shells, the soft or fully annealed temper is used. For the coppers and single-phase alloys, grain size is the basic criterion by which deep drawability is measured. In general, for a given alloy and gauge, ductility increases with grain size and strength decreases. However, when the grain size is so large that there are only a few grains through the thickness of the sheet or strip, both ductility and strength, as measured by tensile testing, decrease. Figure 10–5 illustrates how percent elongation in 2 in. changes with grain size for three different thicknesses of Alloy C26000, cartridge brass.

GRAIN SIZE IN DEEP DRAWING AND PRESSING

For best deep-drawing properties, it is not good practice to specify a grain size larger than that which provides maximum elongation. Low tensile strength is undesirable for any fabricating operation which

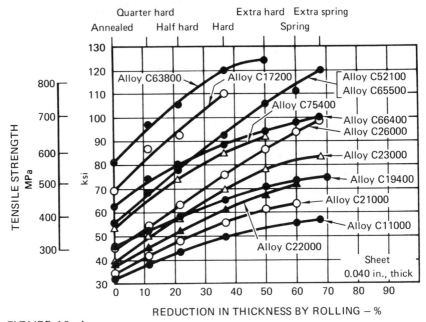

FIGURE 10–4

The increase in tensile strength that occurs during cold rolling is shown for copper and copper alloys. The steepness of the slope for any alloy is indicative of its work hardening rate. (Reference 1).

applies a tensile stress to the metal; hence the average grain size which provides peak elongation should be near the top of the range specified. For the 0.006 in. gauge brass in Figure 10–5 the peak elongation occurs with an average grain size of 0.020 mm. For best deep-drawing properties a range of 0.010 to 0.020 mm average grain size would be most desirable. For the 0.0157 in. gauge brass, a range of 0.030 to 0.050 mm average grain size would provide maximum drawability. For the 0.032 in. gauge brass, a range of 0.060 to 0.090 mm average grain size would give excellent drawability. In fact, this particular gauge and grain size range is frequently used in the production of top and bottom tanks for automobile radiators. These are difficult-to-draw, deep, rectangular shells for which brass is an ideal material because of its deep drawability, corrosion resistance, solderability, and strength.

For forming operations which apply compressive stresses, such as coining or embossing, and for spinning operations, the largest grain size that is compatible with other requirements may be used. In general, however, it is desirable to use metal of the smallest average grain size that can be economically fabricated into the desired part. The surface of the

PERCENT ELONGATION VS GRAIN SIZE FOR ALLOY C26000

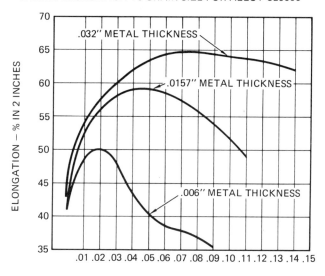

FIGURE 10–5
The relationship between grain size, thickness, and percent elongation for alloy C26000.

part will be smoother, and the part will be stronger and less subject to incidental damage during fabrication, assembly, and service. In addition, there is less chance that failure will be caused by localized overstressing in press drawing.

While the first consideration in selecting the temper or grain size of metal to be deep drawn is that it must be suitable for producing the part without fractures, the fabricating process also can influence the selection of the grain size range. If the part will be fabricated in a series of single draws with an interstage anneal or two, the grain size should be the largest that is compatible with the surface finish required and the thickness of the metal. When surface finish is not important, a relatively large grain size should be selected for heavier gauges. For example, in the deep drawing of brass cartridge cases, Alloy C26000 0.135 in. thick is applied, and a range of 0.070 to 0.110 mm average grain size is used in the production of the initial cup. This cup is subsequently annealed and redrawn to produce the finished cartridge case. In contrast, a 0.072 in. thick Alloy C26000, with a grain size range of 0.025 to 0.040 mm, is used to produce deep-drawn sink strainers. In this case, there is no interstage anneal; additionally, the flange is polished smooth before the part is chrome plated. The tooling must be carefully designed to avoid frac-

(a)

FIGURE 10–6
Two different doorknob designs. Parts such as these require very good formability.

tures with this small grain size, which is chosen to provide strength and a smooth surface for the polishing.

There is often enough difference between the tooling used by separate fabricators to require that a different grain size specification be used by each. The part, the thickness, and the tooling are all important factors in selecting the grain size for a particular alloy. In general, to produce a given part in a multiple-operation press, metal with a smaller grain size can be used than is needed for single-operation pressing. Transfer presses, eyelet machines, and progressive dies are multiple-operation fabricating processes. Heat is generated in pressworking, and when the part becomes warm, the metal flows more readily. This allows a

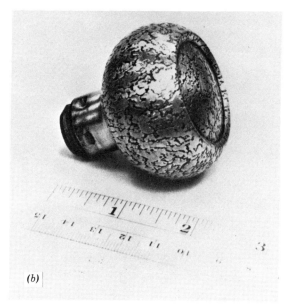

(b)

FIGURE 10–6 *Continued*

smaller grain size to be used, or more cold work to be performed, than in single operations which allow the parts to cool between successive operations.

GRAIN SIZE AND SURFACE

As mentioned above, the surface finish required on the drawn part is an important consideration when selecting the grain size to be used. When metal with a coarse grain size is drawn, the surface roughens and develops an appearance which resembles *orange peel*. Such a surface is more difficult and costly to polish and buff. Therefore, when a part requiring a buffed surface is to be produced in draw presses, much effort is spent in designing the tools and process to use brass with a fine grain size. A classic example of such a situation is the one-piece brass or bronze doorknob. These useful and decorative articles of door hardware are made by the millions. Figures 10–6 *a* and *b* illustrate two of the designs used. Such shapes are difficult to produce on draw presses. These parts are usually produced in transfer presses, and it is not unusual for the process to include between 15 and 20 operations with one interstage anneal or partial anneal. The brass, Alloy C26000, or

FIGURE 10–7
Typical recessed bathroom fixture which requires fine-grained strip to produce a smooth as-drawn finish.

commerical bronze, Alloy C22000, strip from which these parts are made is usually about 0.030 in. thick, and the grain size is usually 0.020 to 0.035 mm or 0.015 to 0.030 mm to provide sufficient ductility for the part to be drawn while avoiding the surface roughening that would occur if a larger grain size were used.

Another classic example of a part that is difficult to draw and requires a fine grain size for ease of finishing is a bathroom recess fixture. An example is shown in Figure 10–7. This is a rectangular part with a rounded bottom requiring a combination of drawing and stretching of the brass from which it is fabricated. Many variations of this part are made, and those which use Alloy C26000 and a thickness of approximately 0.030 in. usually require a grain size range of 0.015 to 0.030 mm or 0.020 to 0.035 mm. These parts are polished, buffed and chrome plated, and the small grain size provides a smooth surface and helps keep down finishing costs.

Table 10–1 gives the ranges of average grain size that are available in high zinc brasses from well-equipped brass mills. Also given are a general statement of the type of drawing operation for which the grain size is usually recommended and a comment concerning the surface characteristics to be expected.

TABLE 10-1
Available Grain Size Ranges and Recommended Applications

Average Grain Size, mm	Type of Press Operation and Surface Characteristics
0.005–0.015	Shallow forming or stamping. Parts will have good strength and very smooth surface. Also used for very thin metal.
0.010–0.025	Stampings and shallow-drawn parts. Parts will have high strength and smooth surface. General use for metal under 0.010 in. thick.
0.015–0.030	Shallow-drawn parts, stampings, and deep-drawn parts that require buffable surfaces. General use for gauges under 0.012 in.
0.020–0.035	This grain size range includes the largest average grain that will produce parts essentially free of "orange peel." For this reason it is used for all sorts of drawn parts produced from brass up to 0.032 in. thick.
0.025–0.040	Brass with 0.040 mm average grain size begins to show some roughening of the surface when severely stretched. Good deep-drawing quality for 0.015 to 0.020 in. gauge range.
0.030–0.050	Drawn parts from 0.015 to 0.025 in. thick brass requiring relatively good surface, or stamped parts requiring no polishing or buffing.
0.040–0.060	Commonly used grain size range for general applications for deep and shallow drawing of parts from brass in 0.020 to 0.040 in. gauges. Moderate "orange peel" may develop on drawn surfaces.
0.050–0.080 0.060–0.090 0.070–0.120	These ranges of large average grain size are used for deep-drawing difficult shapes or deep-drawing parts for gauges 0.040 in. and greater. Drawn parts will have rough surfaces with "orange peel" except where smoothed by ironing.

STRIP SURFACE AND PRESSWORKING

The effects of grain size appear where the metal has been stretched. In these areas the surface characteristics of the original strip tend to be obliterated. In the areas of the part where the metal has been compressed, the strip surface characteristics are accentuated. Where the surface is ironed by tools, it is usually smooth. Brass sheet and strip for application in drawn parts to be polished and plated, or lacquered, are manufactured with a surface as smooth as is compatible with other requirements. It is important to recognize that in this case "smooth" need not mean mirrorlike or reflective; rather it denotes the ability to be easily brought to a high luster by polishing.

In brass mills strip is usually processed between 20 and 30 in. wide. It takes very high roll forces to reduce the thickness of the strip during each pass through a rolling mill. To protect the metal surfaces from damage and reduce the heat buildup, mineral oils with additives and water-soluble oils are used as lubricants. These lubricants must have carefully chosen properties. They are designed to (1) protect the metal during rolling, (2) cool the rolls, (3) be of a composition which will not stain the metal, (4) evaporate from the metal during annealing without leaving harmful deposits, and (5) be readily removable in degreasing operations. When the lubricant is effective, the metal surface will be smooth, matte or lusterless, or lightly burnished, and free of the scratches and scuffs that poorly lubricated metal contains. Some types of rolling require a water-soluble lubricant. This usually produces a more lustrous surface. When a very light water-oil solution is used, the metal is brighter and more reflective. The differences in the surface roughness which are occasioned by the rolling lubricants used are microscopic in scale in terms of the surfaces of the parts produced by drawing and forming. A smooth, dull, nonreflective surface may polish and buff just as readily as a shiny, reflective one after forming.

If well-polished tools are used together with a thin, low-film-strength lubricant, metal with a lustrous surface sometimes can be drawn more readily than metal with a less lustrous surface. The fine asperities which are beneficial in producing uniform coverage by lubricants are left unlubricated by a thin and watery solution, and excessive friction can result. When a lubricant with good film strength is used, either the burnished, lustrous surface or the unburnished, matte surface will be protected, but the duller appearing metal will be less likely to be damaged by lack of lubricant coverage. Generally, the parts fabricator can more readily adjust the lubricant used in a pressworking operation than a brass mill can on a rolling mill, where hundreds of gallons of lubricant are involved. When a pressworking problem due to incompatibility of lubricant and metal surface is encountered, it is normally solved by adjusting the press lubricant.

Successful pressworking is highly dependent on careful control of the frictional forces between the tools and the metal. The friction between the metal and the hold-down pad and the metal and the die provides the tension which draws the metal to the shape of the punch. The gradual and uniform release of the metal under the hold-down pad allows it to flow into the die without developing wrinkles and puckers. The thickness of the metal, clearance between tools, condition of the metal surfaces, and lubrication all help to determine how high and how uniform the frictional forces will be.

Each alloy has inherently different coefficients of friction against tool steels. For example, the coefficient of friction of copper against tool steels is higher than that of brass. The high zinc brasses have coefficients of friction lower than the copper-rich brasses, and these values vary with zinc content. The phosphor bronzes have high coefficients of friction compared to high zinc brasses. The nickel silvers have higher coefficients than brasses, and cupro-nickels higher values than the nickel silvers. While values for coefficients of friction of copper and the copper alloys are given in the literature, they are meaningful only for a particular test situation. The general statements given above result from practical experience and will serve to explain differences which are sometimes observed between the responses of various alloys drawn into similar parts.

As well as differences in surface occasioned by rolling practice, differences in surface can also result from combinations of annealing and cleaning (Refer to Chapter 2 on melting and mill processing). Metal for draw-press operations is usually furnished in soft or annealed tempers and has been subject to some combination of annealing and cleaning practices. When large grain sizes are required, the high annealing temperatures accentuate surface reactions in the furnace. Batch annealing may have been done in furnaces with controlled atmospheres to prevent oxidation, or in partially controlled atmospheres resulting from products of combustion of gas used to heat the annealing furnace, or in essentially open atmospheres where surface oxidation can be heavy. The type of control exercised to avoid oxidation often depends on the alloy. When the annealed surface differences are combined with a variety of cleaning processes, the result can be a great many slightly different surface conditions in the as-shipped product.

Unless specifically instructed to degrease, most mills will lubricate the metal in some manner to prevent friction damage during transit and uncoiling in the user's plant. This lubrication can be an aid in drawing or forming parts. Since lubrication is such an important part of successful presswork, however, it is always good practice to employ lubrication at the press itself. This will insure that any small variations in surface condition will not affect the operation.

There are surface conditions that are decidedly detrimental to drawing operations, and the parts fabricator is expected never to use defective metal. Metal can become stained because of improper processing such as failure to dry it thoroughly after cleaning. Moisture stains can originate in shipping or can occur in storage. Metal received on a cold day and placed in a warm, moist storage area will quickly pick up condensate and become stained unless unpacked and dried by circulating air through it. Stained metal should not be used in draw presses. It causes tools to seize; malformed or broken parts result. Other surface conditions, such as mechanical defects like scratches and gouges, may originate in the mill or during the fabricator's handling and uncoiling. Scratches and gouges which have been rolled over can be recognized as originating in the mill. The same is true of slivers and blisters. Such mill defects that interfere with parts fabrication should be brought immediately to the supplier's attention, so corrective action can be taken.

DRAWING OF METAL IN PRESSES

With an understanding of the influence of the inherent characteristics of copper and copper alloys on drawability, a review of *draw-press* and *forming operations* themselves and the factors important to their success and failure is in order.

The drawing of metal in presses is a process by which flat metal blanks are formed into various hollow shapes. Drawing tools generally consist of a punch, die, and pressure pad, as illustrated in Figure 10–8. The most significant dimensions of the original blank and the drawn part are indicated in Figure 10–9.

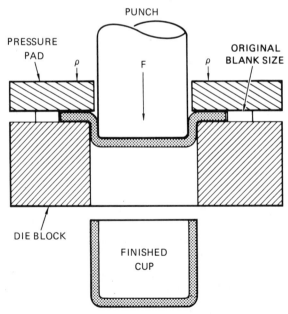

FIGURE 10–8
Schematic drawing showing the basic elements of a draw press operation. (Doyle et al. [6], p. 285)

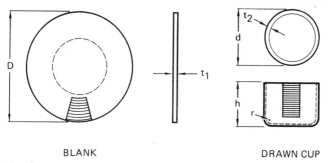

BLANK   DRAWN CUP

FIGURE 10–9

Significant dimensions of drawn parts: D, initial blank diameter; t_1, blank thickness; d, drawn cup outside diameter; t_2, sidewall thickness; h, cup height; r, bottom cup radius. (Doyle et al. [6], p. 285)

The blank is centered over the die and held between the pressure pad and the die block. As the punch encounters the blank and pushes it into the draw die, the metal is quickly stressed beyond its elastic limit and begins to yield. The metal flows plastically in the direction of draw in both tension and compression: in tension, in the direction of draw; in compression, around the circumferential flow line. As the metal is forced to flow between the pressure pad and the die face, there is an increase in thickness and a decrease in surface area. Hold-down pressure is applied to the blank by the pressure pad as the metal flows toward the die radius. This prevents the flange from wrinkling as the metal is compressed and drawn into the die. The bottom of the drawn shell is changed very little in hardness, while the sidewalls are work hardened markedly. The amount of hardness increase depends upon the work-hardening rate of the alloy and the percent of area reduction from blank to drawn shell.

Drawing a rectangular shell results in varying degrees of metal flow in different areas. The corners are severely worked, but the sides and ends are not. Drawing occurs in the corners, while the sides and ends are formed by bending and stretching. In the corners the metal flow is in compression as it moves toward the die radius and in tension as it flows over the radius.

To insure proper flow of metal, good tool design practices must be followed. The edges of punches and dies are rounded to avoid tearing or cutting. These radii are polished and well blended into the sides of the tools so the metal flows over them smoothly. A reduction in the amount of friction or drag can be provided by designing a short land and relief angle on the die face. This reduces the surface area which is in contact with the workpiece during drawing.

TYPES OF DRAWING OPERATIONS

Drawing operations are sometimes subdivided into four types, as follows:

1. *Shallow draw*—usually describes a drawn part with a depth less than 50% of the cup diameter.
2. *Deep draw*—describes a drawn part whose depth is over 50% of the cup diameter; usually one or more redraws are used.
3. *Redraw*—describes a press operation for further diameter reduction after the initial draw or cupping operation. The workpiece is a cup or a drawn shell. Some drawn parts may require a number of redraws to attain the final depth and sidewall dimensions.
4. *Reverse redrawing*—the cupped workpiece is redrawn in the opposite direction to that of the initial drawing operation, turning the shell inside out. Reverse redrawing is used instead of direct redrawing:
   (a) To obtain better distribution of metal in a complex part.
   (b) To make greater reductions per redraw to eliminate intermediate anneals.

OTHER DRAW-PRESS OPERATIONS

Some other special draw-press operations include the following.

IRONING. A drawing technique that reduces the wall thickness of a shell (Figure 10–10) to the thickness desired, which is the clearance between punch and die. Ironing is done for several reasons:

1. To counteract the normal tendency of a drawn part to thicken in the sidewall during drawing. (Tool clearances can be set at minimum stock thickness, and the sidewalls of the finished part will be uniform.)
2. To reduce the wall thickness.
3. To add strength by work-hardening the sidewall.
4. To increase the length of the shell, thus requiring a smaller blank size.
5. To obtain a burnished surface.

In ironing, a certain amount of diameter reduction is done to allow for free entry of the punch. The reactive residual stresses left in the shell are compressive. If external machining is done later, there is less chance of cracking and distortion than there would be with an unironed drawn shell.

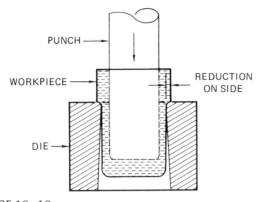

FIGURE 10–10
Schematic drawing of an ironing operation. (Doyle et al. [6], p. 300)

BULGING.　Producing an irregular contour along the length of a shell by expanding the diameter for a portion of the length (Figure 10–11). This is accomplished by the use of rubber punches, fluids, segmented dies, a wedge action punch or die, or a combination of these which provides an expanding action when the ram pressure is applied.

SIZING.　A secondary press operation used to set a specific dimension in a shell. It is usually done by confining the metal and forcing it to flow

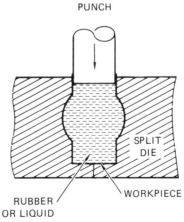

FIGURE 10–11
Schematic drawing of a bulging operation.
(Doyle et al. [6], p. 293)

by squeezing or crowding. Stretch sizing is another technique which may be used to eliminate oil canning in the sidewall of a rectangular shape or to stretch the shell to establish a particular length.

NECKING. A diameter-reducing operation on the mouth of a shell. It can be accomplished by spinning, by swaging, or by reducing in tapered press dies (Figure 10–12). As the diameter is reduced, the metal is compressed and wall thickness increases.

MISCELLANEOUS FORMING OPERATIONS

Some forming operations closely related to drawing are the following.

COINING. A cold pressworking process by which the workpiece is confined and struck or embossed to conform to the configuration of the dies (Figure 10–13). It can also be the restriking of a drawn or stamped part to sharpen a dimension, harden, or form a profile on a part. Coining can provide dimentional accuracy equal to that obtained by machining without the added cost.

The single-phase copper alloys have excellent coinability. However, the alloys with higher compressive yield strengths require stronger,

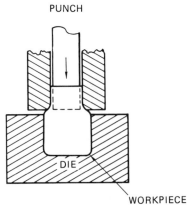

FIGURE 10–12
Schematic drawing of a necking operation. (Doyle et al. [6], p. 293)

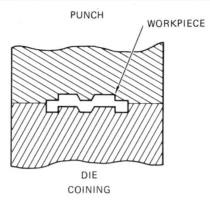

FIGURE 10–13
Schematic drawing of a coining operation. (Doyle et al. [6], p. 300)

more powerful presses. Pure copper is easy to coin. The single-phase copper-zinc alloys are all coinable, but decreases in ease of coinability with increase in zinc content. The cupro-nickel and nickel-silver alloys are used for coins that are silver in color. The amount of metal movement that can be accomplished in one coining operation will depend greatly on the unit load in compression that the coining dies can withstand without deforming.

When coining medallions or tokens, it is good practice to blank the metal in the hard condition and anneal and clean the blanks prior to coining. This will insure a relatively flat bank, minimize the amount of burr present from the blanking operation, and provide soft edges on the blank, so the metal will flow readily to fill the die cavity. The coined surfaces will mirror the surface quality of the die, and defects can be minimized by keeping the dies highly polished and clean—free from dirt, slivers, and excessive lubricant. Tool alignment is important for obtaining dimensional accuracy in the coined part. Tool breakage is always a potential problem because of the shock forces applied in coining. When successive coining operations are used, introduction of an anneal in the process can help to avoid tool breakage.

STRETCH FORMING. The forming of sheet metal over a block made to the contour of the finished part (Figure 10–14). The metal is held in tension and stretched beyond the yield point, usually 2 to 4% to set it in the shape of the part. *Stretch draw forming* uses both stretch forming and conventional press forming equipment (Figure 10–15).

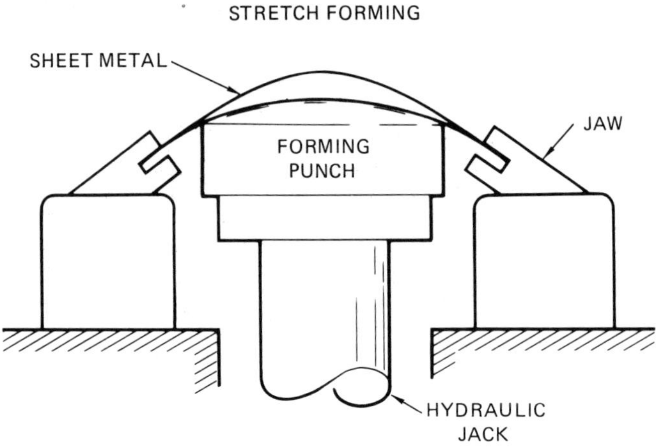

STRETCH FORMING

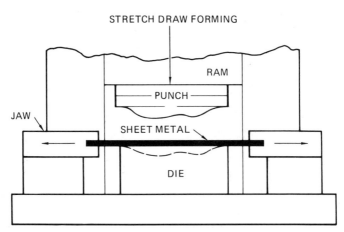

FIGURE 10–14
Schematic drawing of stretch forming. (Doyle et al. [6], p. 292)

The advantages over conventional press forming are:

1. Less force is required to form the part.
2. There is less chance of buckles and wrinkles occurring in unsupported area of the part.
3. Springback is less.
4. Residual stresses are lower.

STRETCH DRAW FORMING

FIGURE 10–15
Schematic drawing of stretch draw forming. (Doyle et al. [6], p. 292)

The disadvantages are:

1. Sharp radii and contours cannot be produced.
2. Irregularities in the surface of the part cannot be ironed out.
3. The process may be slower than competitive processes.

When stretch forming, care must be taken not to overstretch the workpiece. This will result in localized thinning in the sidewall. Any foreign material, such as metal slivers picked up by the forming block, will show up as indentations on the part surface. A clean lubricant and clean tool surface must be maintained to avoid surface defects.

BLANKING. The cutting of flat shapes or blanks from metal in a blanking die (Figure 10–16). The metal is stressed in shear beyond its ultimate strength until fracture occurs. Tool clearances are important for producing a blank having minimum edge distortion and free of burrs. On thicknesses up to 0.150 in. a 5 to 10% clearance between the punch and die on a side is good practice; above 0.150 in. clearance should be closer to 10 to 12%. To blank annealed temper stock cleanly requires less clearance than to blank harder tempers; however, distortion of the blank edges is more prevalent in blanks made of softer material. Harder tempers of an alloy require greater shearing forces but produce a cleaner shear and flatter blank. If the clearance is too little, tool edge wear and galling will result.

A blank will be the size of the die or female member of the blanking tools. The clearance should be ground off the punch to produce a blank

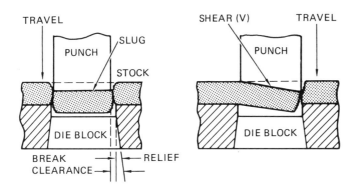

(A) CONVENTIONAL CUTTING   (B) PUNCHING WITH SHEAR

FIGURE 10–16
Schematic drawing of blanking operations. (Doyle et al. [6], p. 277)

of accurate size. In Figure 10–16*B*, shear is shown on the punch, and this is good practice when piercing holes in a part. However, when shear is used to reduce blanking force requirements, the shear should be ground on the die face to avoid distortion of the blank.

SPINNING. A process of hand-shaping a shell. The workpiece is clamped to a revolving form block or chuck in a lathe and forced to conform to the shape of the forming block (Figure 10–17.) This is achieved by pressing a smooth tool against the surface of the workpiece in a series of sweeps until the final contour is achieved. A part that can be drawn also can be spun. When producing small quantities of large shells, spinning is often cheaper than drawing. Spinning tools are inexpensive compared to press tools, and set-up time is short. For long production runs, however, press forming is superior. Spinning is a slow process.

The coppers and brasses are well suited for spining. Most brass musical instruments are made from spun parts assembled by brazing.

HOLE FLANGING OR DRIFTING. The turning up or extruding out of the edge of a pierced hole (Figure 10–18). This is usually done to strengthen by increasing the restraining area around the hole, or to

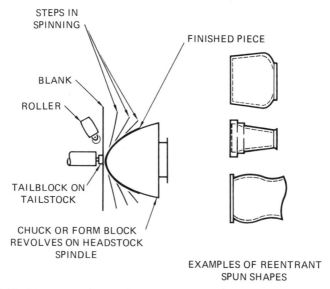

FIGURE 10–17
Schematic drawing of the spinning operation and spun parts. (Doyle et al. [6], p. 290)

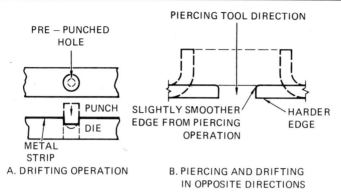

PIERCING TOOL DIRECTION

PRE – PUNCHED
HOLE

PUNCH
DIE

METAL
STRIP

A. DRIFTING OPERATION

SLIGHTLY SMOOTHER
EDGE FROM PIERCING
OPERATION

HARDER
EDGE

B. PIERCING AND DRIFTING
IN OPPOSITE DIRECTIONS

(MANUFACTURING PROCESSES AND
MATERIALS FOR ENGINEER (7) P. 293)

FIGURE 10–18
A. Schematic drawing of drifting operation. B. Drifting is less likely to cause edge tearing if done in the direction opposite to piercing. (Doyle et al. [6], p. 293)

create a collar on the hole for a thread tapping operation. Determination of the length of drift that can be made in a particular material is a trial and error process. Drifting can be performed on most copper alloys if the temper used is appropriate. Best results are obtained by piercing the hole in the direction opposite that of the drift operation. This practice places the smoothest edge of the hole in the position which will be stretched the most, and reduces the tendency for edge tears in the drift.

BRASS BLANK, CUP, AND REDRAWN SIZES

Selection of the best material for a drawn part requires consideration of many factors. A detailed analysis of the required properties of the final part is the primary step. Once such property requirements as corrosion resistance, strength, formability, and color are determined, consideration of the part fabrication process can proceed.

The calculation of blank size and the design of the steps required for the fabrication of a particular part are beyond the scope of this work. There are tool engineering and design texts which cover these subjects that the interested reader can consult. Some rules of thumb for the drawing of brass parts are given below so the reader will gain a general idea of the approach which is taken.

The blank size is calculated based on the part dimensions. As a general rule for brass, a 40 to 45% reduction of the blank diameter can be accomplished in the cupping operation. The resulting cup diameter will be 55 to 60% of the original blank diameter. The percent reduction is determined by the forumla:

$$\text{Percent Diameter Reduction} = 100 \left(1 - \frac{\text{Cup O.D.}}{\text{Blank O.D.}}\right)$$

Redraw operations on brass cups or drawn shells usually are designed to produce a reduction in diameter approaching 20% or an area reduction in the range of 25 to 35%. Area reduction is calculated from the following formula:

$$\text{Percent Area Reduction} = 100 \left(1 - \frac{t_2 \times \text{O.D.}_2}{t_1 \times \text{O.D.}_1}\right)$$

where t_2 = sidewall thickness after redraw
O.D._2 = outside diameter after redraw
t_1 = sidewall thickness before redraw
O.D._1 = outside diameter before redraw

Area reduction can also be used to estimate the strength that will be imparted to the drawn shell by the cold reduction. The total percent of area reduction between the blank and the drawn shell can be considered to have a work-hardening effect equivalent to a gauge reduction by rolling of an equal percentage. Tabel 10–2 lists the cold-rolled tempers and the rolling reductions needed to obtain them. This table is a guide to the temper equivalent produced by drawing a part.

TABLE 10-2
Cold-Rolled Temper versus Reduction in Thickness

Temper Name	Reduction, %
¼ Hard	10.9
½ Hard	20.7
¾ Hard	29.5
Hard	37.1
	44.0
Extra Hard	50.0
	55.7
Spring	60.5
	64.8
Extra Spring	68.7
	72.0
	75.2

LUBRICATION

Earlier in this chapter some of the variables that influence the surface characteristics of copper base metals were discussed. In drawing parts from sheet metal a surface which provides uniform and low friction with tooling is most important. To achieve this, good lubrication is required. Inadequate lubricants are the most frequent source of trouble. Proper choice of a good lubricant can overcome shortcomings in the design of punch and die, inconsistency of metal surface, and severity of draw. The proper lubricant should be capable of providing a film on the metal surface which is a friction reducing cushion between the metal and the drawing tools.

Proper lubrication is essential to provide:

1. Even flow of metal between punch and die.
2. Low tool wear rate.
3. Cooling of tools to reduce dimensional changes during extended use.

Typical indications of insufficiency are galling of parts, loading of tools, fracturing, and uneven top cups.

The two basic kinds of lubricants used in draw press work are water-base and oil-base. *Water-base* types are often used in high speed press operations because of their effectiveness as coolants. *Oil-base* lubricants, on the other hand, may be desirable to protect the workpiece from stains during prolonged storage. The choice depends upon the severity of the draw and also on secondary considerations such as staining, versatility, and ease of application and removal.

WATER-BASE LUBRICANTS. A list of water-dispersible types is given below. The soap-plus-fat compounds of this group are the most versatile. Consistency can be readily changed by additions of pigment or water. In severe draws on heavy-gauge material pigment additions are most helpful, providing a soft, solid cushion between metal and tools. Water-soluble oils (also water-base) provide the best cooling for tools and workpiece. Since they are less effective in avoiding galling or scoring than soap-plus-fat compounds, they are best used in low-forming-pressure work. Soap solutions can be readily removed by hot water and lend themselves to easy cleaning. Dry soap is primarily helpful in deep-drawing brass shells and heavy gauge copper alloys.

1. *Soluble Oil*—This forms a milky-opaque emulsion when added to water. It is prepared by adding emulsifiers to straight mineral oil.

Heavy-duty types contain sulfurized or chlorinated compounds. Usual formulas range between 1 part oil to five parts water and 1 part oil to 20 parts water.

2. *Soap Solution*—This is a concentrated solution of potassium or sodium soap in water. It is rapidly soluble and self-cleansing. Slip qualities are excellent; uniform wetting provides unbroken film. Foaming is a drawback.

3. *Nonpigmented Soap-Fat Compound*—This consists of 35 to 60% fat, plus soap and water. Product resembles mayonnaise. Performance is highly rated. Most popular for moderately severe draws. Usual mix ratios are between 1 to 1 and 1 to 5.

4. *Pigmented Soap-Fat Compound*—This contains pigment concentrations of 10 to 50% in soap, fat, and water. Made for severe drawing. Serious shortcoming is difficult removal.

5. *Dry-Film Soap*—This is made principally of sodium soap and a water-soluble spacing agent, such as borax. Parts are immersed in a 10 to 25% solution at 180 to 200°F for 3 minutes and air dried. Thin, dry, plastic film lubricates much better than aqueous solutions. Other advantages: less mess, easy cleaning and handling. Drawbacks: requires special dipping and drying equipment; high humidty adversely affects the film.

OIL-BASE LUBRICANTS. A list of oil type lubricants is given below. Volatile solvents are used for mild forming operations and may eliminate the need for cleaning. Medium-severe draws can use mineral oils of low to high viscosity, depending on the part. Severe draws will require fatty or chlorinated oils.

6. *Straight Mineral Oil*—Any naphthenic type is satisfactory. Its viscosity must be between 100 and 300 Saybolt seconds at 100°F. Chemical stablity is not necessary, nor is minimum-viscosity-temperature change of high paraffin grades. Advantage: low cost.

7. *Mineral Oil, Oily Blend*—This is a mixture of straight mineral oil and 5 to 50% oily materials. Fatty oils (polar compounds) are added to combat high pressures which break lubricant film of straight mineral oils; also, a little more expensive.

8. *Straight Fatty Oil*—Lard oil is most popular. Combats exceptionally high drawing pressures. Disadvantage: more costly than straight mineral oil.

9. *Sulfurized Mineral Oil Blend (noncorrosive)*—This is made by mixing a mineral oil and from 5 to 50% noncorrosive sulfur compounds.

Usually found in draws of copper, brass, or bronze, it is less desirable than fatty oil because of its odor and color.

10. *Sulfurized Mineral Oil Blend (corrosive)*—This is the same as 9 except that the sulfur is quite active. Used extensively on carbon steels for extreme pressure applications. Not as expensive as 7 to 8.

11. *Chlorinated Mineral Oil Blend*—This is made by mixing mineral oil with 5 to 50% chlorinated oils. This lubricant is often more satisfactory than sulfurized blends and more universal in application. Cost is comparable.

12. *Straight Sulfurized Oils*—These complex compounds are made by reacting sulfur chemically with mineral oil, fatty oils, fats, fat derivatives, or terpenes. They are among the most expensive lubricants. Used for extreme pressure applications.

13. *Straight Chlorinated Oils*—These compounds are gaining popularity over straight sulfurized oils. Light colored, nonvolatile, and syrupy, they are made from the reaction products of chlorine, hydrocarbons, fats, or waxes. Cost is about the same as sulfurized oils. Also used for extreme pressure applications.

A guide to the use of water-dispersible and oil type lubricants for various kinds of presswork with different metals is given in Table 10–3.

TABLE 10-3
Guide to Selection of Lubricants for Presswork

Operation	Copper, Brass, Bronze	Aluminum	Carbon Steel	Alloy Steel	Stainless Steel
Stamping	1, 2, 3	2	1, 2, 3	3	4
	6, 7, 9	6, 7	6, 7, 10	10, 11	11
Shallow draws	2, 3	—	3	4	—
	7, 9	7, 8	7, 10, 11	13	11, 13
Deep draws	3, 4	—	4, 5	4, 5	—
	8, 11, 12	8, 11	11	11	11, 13

Other considerations besides draw severity sould be weighed in selecting lubricants. Coppers, for instance, have a marked tendency toward cold welding, and filtering is important to prevent slivers from contaminating the lubricant and marring the surface of the part. Sulfur or chloride compounds can form stains if not cleaned quickly. Additional points of importance are the following.

STABILITY IN STORAGE. Since lubricants are made up several months ahead, they must keep well. Liquids should not gel or separate; solids must not bleed oil or form crusts.

Inhibitors are needed to prevent rancidity in fats. Pigmented, soap-fat pastes must have sufficient body to prevent separation of dispersed solids.

EASE OF APPLICATION. This feature is important to operators who work on incentive. A convenient method is continuous circulation. Lubricants should be nonfoaming. Other methods involve coated rollers, dipping, spraying, and drip systems.

EASE OF REMOVAL. Water or solvent degreasers are the most common cleaning agents. Choose the one that matches the lubricant.

Some plants limit the selection of drawing lubricants to those most easily cleaned or those which can be removed by equipment on hand. This can be a cart-before-the-horse procedure which ends up being more expensive than improved cleaning equipment.

Welding operations, which follow drawing, affect the choice of lubricant. Certain water-dispersibles are most suitable.

OVERALL ECONOMY. Obviously the cheapest lubricant is not always the best. Be sure to consider production labor cost; reconditioning, tool life, cost of metal, processing of rejects, press overhead, and cleaning.

UNIVERSAL FACTORS. Any compound will work on some jobs. On other jobs every compound fails. Such behavior can be explained by:

1. Type of materials.
2. Their condition.
3. Severity of the draw.
4. Die efficiency.

Materials (tools and workpiece) are important. Hard tools resist welding better than soft ones. Tungsten carbide or chrome plating withstands metallic buildup longer than high carbon steels. Mirror finished tools (diamond lapped) appreciably extend service.

Poor surface condition greatly increases drawing force. Abrasive oxides, dirt, scale, and hydrogen embrittlement ought to be avoided.

Surfaces should be physically and nearly chemically clean and slightly rough to retain the lubricant.

The drawability of metals varies considerably. High-yield-strength materials require greater forces for plastic deformation than those of lesser strengths, leading to greater pressures on the lubricant. In general, brass is easier to draw than carbon or stainless steels.

Die efficiency is often overlooked. Construction, clearances, and angles must conform to the metal limitations. Sharp corners cause localized, heavy drawing pressures, requiring heavy drawing pressures, requiring heavy drawing oils.

Often, simply correcting the die design will cut rejects and may eliminate the need for a costly lubricant.

DRAWING DEFECTS—DESCRIPTION AND CORRECTION

1. *Excessive earing*—Four evenly spaced ears appear on a brass cup; the problem is usually one of directionality in the material. Directionality can be minimized in the processing of the strip. With knowledge of the part to be produced, the brass mill will set up the process to produce metal with the least directionality. See Figure 10–19.
2. *Earing at One Point on Periphery*—This condition is usually caused by nicks, excessive burr, or other imperfections on the periphery of the blank. Stained spots on the metal or uneven distribution of lubricant can also cause this condition. See Figure 10–20.
3. *Out-of-Round Condition*—Too much clearance in the tools will allow the cup to take an irregular shape. Tightening up these clearances

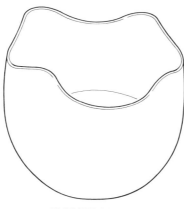

FIGURE 10–19
Excessive earing.

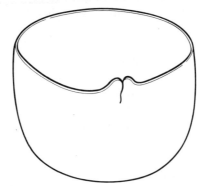

FIGURE 10–20
Earing at one point on periphery, caused by
nicked blank edge.

or slight wall ironing will improve this situation. The tendency for
out-of-roundness varies with thickness of the metal and diameter.
4. *Wrinkled Sidewall or Flange*—The hold-down pressure is not high
   enough to contain the material. Increasing the pressure or increasing the blank size is a possible corrective measure. See Figure 10–21.
5. *Fractured or Broken-Out Bottoms*—This can result from too much
   hold-down pressure, too much draw reduction, or poor lubrication.
   The basic problem is that the stress applied exceeds the ultimate

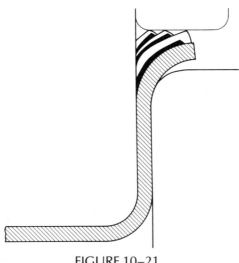

FIGURE 10–21
Wrinkled flange.

strength of the material. Consideration should be given to all means that will reduce the stress.

6. *Uneven Wall Thickness or Variation*—Stained material can cause this condition because it will not allow the material to flow evenly. A misalignment of tools may also be the culprit if the wall thickness varies from light to heavy 180° apart.

7. *Excessive Thinning in the Bottom, Bottom Radius, or Sidewall Adjacent to the Bottom*—There is a natural tendency for the metal to thin out in these areas. However, if metal flow over the die radius is restricted, the pad pressure is too great, or a rounded bottom punch is used, the thinning can be excessive. The natural tendency to thin can be overcome by roughening the punch with a stone or machining rings around the radius of the punch. The metal will then be prevented from flowing over the punch radius and will have less tendency to thin.

8. *Scratches or Score Marks in the Sidewall*—These usually mean galling, wearing, or metal pickup on the tools. Improved lubrication and polishing of tools should correct.

9. *Excessive "Orange Peel"*—When roughening of the surface occurs in a stretched area of the part, large grain size is the usual cause. The use of metal with a finer grain should minimize this defect.

10. *Uneven Flange*—This usually is accompanied by uneven wall thicknesses due to punch and die misalignment. It can also be the result of improper nesting of the workpiece. See Figure 10–22.

11. *Excessive Tool Wear*—Stains on the surface, especially on silicon or phosphor bronzes, can wear tools. Generally, this is not a problem with other copper alloys. Whenever metal or parts are annealed, a protective atmosphere is desirable and thorough cleaning and drying will prevent stains.

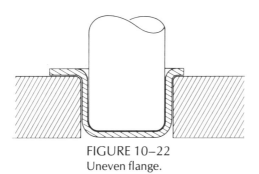

FIGURE 10–22
Uneven flange.

FORMING BY BENDING

A great many parts which are fabricated from copper and copper alloys are formed by what are essentially bending operations. These parts are frequently ones which perform springlike functions and are fabricated from the higher strength rolled tempers. Chapter 13 on designing springs is concerned with the functions of such parts. In this chapter some characteristics of the metals that influence their bendability will be considered. Some generalizations are (1) the harder the temper of rolled metal, the larger is the radius over which it must be bent to avoid fractures; (2) the higher the strength level of an alloy, the larger is the bend radius needed; (3) bends made across the direction of rolling (the good way) can be made over smaller radii than bends parallel to the direction of rolling (the bad way) without fracturing; (4) thicker metal must be bent over a larger radius to avoid fractures than thinner metal of the same temper and alloy; (5)not strength per se, but rather the alloy and the amount of cold work required to attain the strength, is the governing factor.

When selecting metal to perform a spring function, the bendability requirement is often one of the limiting conditions. For most applications, the smallest possible bend radius is usually preferred. Therefore a combination of high strength and good bend formability is most desirable. Since bends perpendicular to the direction of rolling can be made over sharper radii, where maximum strength is required part design and tool layout often take this factor into consideration. Bends made in any direction other than parallel to the direction of rolling or 90° to the direction of rolling will have minimum bend radius requirements between those of the 0° and 90° directions. To utilize strip most economically part blanks whose shape is appropriate are nested at 45° to the direction of rolling. Figure 13–8 illustrates what is meant by direction of bending. Figures 15–3 and 15–4 illustrate how blank layout direction affects bending and strip utilization. Chapters 13 and 15 both cover some aspects of bending direction.

Frequently used bend data define the minimum bend radius that will avoid fractures when bending various thickness-temper combinations to a 90° angle in the 0° or 90° direction relative to the direction of rolling. Figures 10–23 and 10–24 illustrate bend data for Alloy C51000, phosphor bronze. For each temper or tensile strength, thickness is plotted against the minimum bend radius that can be used without fractures occurring. Any bend radius larger than the minimum should produce bends without fractures. All bends are 90° bends, perpendicular to the rolling direction in Figure 10–23 and parallel to the rolling direction in Figure 10–24. These data are from laboratory bend tests,

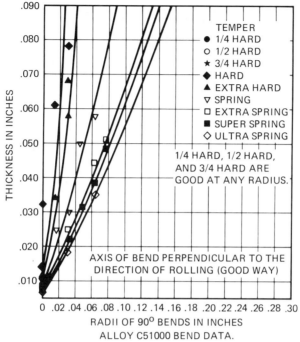

FIGURE 10-23
Bend data for Alloy C51000 for bends across the direction of rolling.

and bends made by parts-fabricating tools will not always be exactly comparable.

As is true of other press forming operations, there are various tooling methods and techniques for bending. No attempt will be made to cover them here. The copper alloys generally have moduli of elasticity in the range of 15 to 20 million psi. For a given amount of displacement they will spring back about twice as far as steel of comparable thickness and strength. Provision needs to be made in designing tools to overbend in some manner to obtain the desired angle of bend. In considering how sharp the bend radius can be, this should be taken into account. Bending techniques which stretch the metal excessively during bending will cause fractures even though the bend radius is not too sharp. Bends made by tools with a wiping action will fracture if proper tool clearance and lubrication are neglected. Scored parts and loaded tools are sure signs of such poor practices.

A theoretical treatment of the variables that affect bend formability follows to explain some of the observed differences between materials.

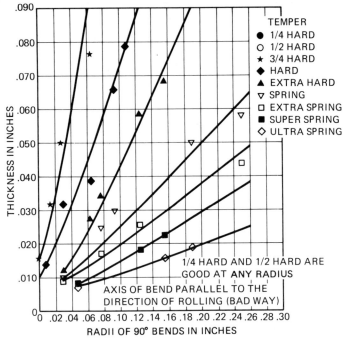

FIGURE 10–24
Bend data for Alloy C51000 for bends parallel to the direction of rolling.

The variables that affect bend formability are ductility, texture, and ability to distribute strain.

Figure 10–25 is a schematic of a bend. To make the bend, a localized area is deformed around a radius (the bend radius). On either side of that deforming area the metal strip remains undeformed. The outer fibers of the bend are in tension, and the inner fibers of the bend are in compression. When forming occurs over tighter bend radii, strain increases at the outer fiber until at some radius bend fracture is noted. Fracture is defined as a visible crack in the outer fiber. This is, however, a subjective determination, and it is not unususal for laboratories (or people within the same laboratory) to disagree as to whether fracture has occurred.

Fracture initiates in bending when a critical strain (E_b) is reached at the outer fibers of the bend. This critical strain, which is constant for any alloy chemistry and processing history, is reduced by cold rolling; ductility is consumed. Qualitatively, this accounts for the temper effects seen in Figures 10–23 and 10–24. Increasing the temper increases the minimum bend radius for a given thickness.

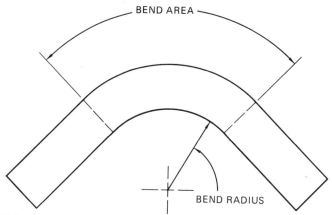

FIGURE 10–25

Schematic illustration of a bend, showing localized area of deformation over the bend radius bound by undeformed regions of the sheet.

In order to study the effect of thickness on the shape of the bend curve, the strain at the outer fiber of the bend is plotted as a function of bend radius for two sheet thicknesses in Figure 10–26. As thickness is reduced for a given bend radius, the strain developed in the outer fiber decreases. The relationship between minimum bend radius and thickness, and the strain developed in the outer fiber, is given by the equation

$$R/t = 1/(1 + E_b)^2 - 1$$

where R = bend radius at fracture

t = thickness

E_b = critical strain at the outer fiber in bending

This equation is derived from the geometry of a bend. For a given alloy chemistry and processing history, E_b is defined and, therefore R/t is a constant. The slope of the bend curves (Figures 10–23 and 10–24) is R/t; each temper has a characteristic R/t.

At this point, we have developed the general shape of the bend curves and have explained the thickness and temper dependence. We will now examine the relationship between bending and tension. The most obvious difference is the gauge length over which each deformation process occurs. In bending the deformation occurs over a very short gauge length, while in tension the gauge length is typically at least 2 in. The lack of recognition of the importance of gauge length upon measured ductility explains the lack of correlation between bend performance and the 2 in. gauge length tensile elongation normally reported for copper alloy strip. In fact, we believe that for most

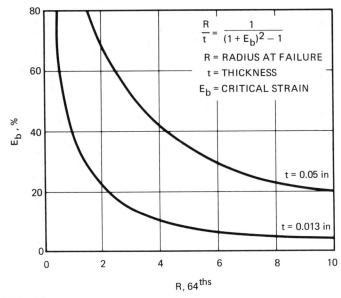

FIGURE 10–26

Outer fiber bend strain plotted against bend radius for two sheet thicknesses. These curves conform to the equation noted above.

conditions tensile elongation and bend ductility are comparable when both are measured at the same gauge length.

An important difference between bending and tension is the ability of the bend to make use of both necking elongation and uniform elongation. Another significant difference between bending and tension is the deformation stress state. In order to consider this, we have to look at the bend in more detail and especially at metal flow and constraint during bending. Figure 10–27 shows a bend characterized by a short gauge length region, where deformation occurs, bounded by two undeformed areas. During bending, a circumferential tension is developed on the outer fibers as they extend. This tension causes the bend specimen to want to thin in the radial direction and narrow in the transverse direction at the same time. The radial thinning is permitted, but the transverse narrowing is not because of the constraint of the elastic material surrounding the bend. Since this transverse (lateral) flow wants to occur but is prevented, a tensile stress is set up in the transverse direction. As a result, the stress state in bending is more complex than in tension.

Since strain occurs only in the radial and circumferential plane (there is no strain in the transverse direction), this stress state is called plane

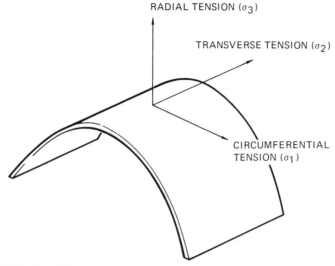

FIGURE 10–27
Schematic illustration of bend region, defining direction of principal stresses.

strain. At the edges of the bend specimen, however, there is less constraint and the stress state approaches uniaxial tension.

Stress state affects ductility as shown in Figure 10–28 for bending, where the strain to failure at the outer fibers of the bend is plotted as a function of the width to thickness ratio. For a constant thickness, one can consider this a fracture strain versus specimen width curve. The geometry factor W/T is important because it influences constraint in

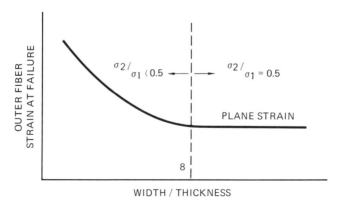

FIGURE 10–28
Plot of outer fiber strain at fracture versus width to thickness ratio.

bending. For width to thickness ratios greater than about 8 to 1, plane strain develops and ductility to fracture is reduced.

Figure 10–29 shows data for Alloy C26000 as a bar graph, plotting outer fiber strain for two width to thickness ratios—one in the plane strain region, and one out of the plane strain region. Ductility is lower in plane strain.

Because of plane strain, the fracture site for most bends is at the center of the bend specimen. Edge fractures and a $W/T>8$ usually indicate a shearing or blanking problem, causing premature fracture.

In order to understand why ductility is reduced in plane strain, we have to look at the yield locus, shown in Figure 10–30. The definition of a yield locus is that it represents the boundary between elastic and plastic deformation. Combinations of stresses within the yield locus are elastic. Stresses outside the yield locus are plastic. For example, consider the case of uniaxial tension. As we begin to increase the tensile stress, we remain elastic until the point σ_A is reached, when plastic deformation occurs; σ_A is the uniaxial tensile yield strength. If we consider plane strain, we have a more complex stress state as stresses are present in two directions. If we start on the yield locus at the point σ_A and increase the transverse stress, we find ourselves back inside the yield locus so that we are elastic. In order for plastic deformation to occur, we then have to increase the circumferential tensile stress to a point σ_B, where we are

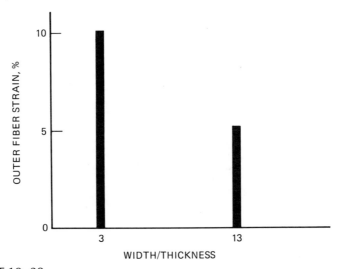

FIGURE 10–29
Laboratory data showing outer fiber strain to fracture versus width to thickness ratio for Alloy C26000, cartridge brass.

MODEL OF BEND FORMABILITY

YIELD LOCUS

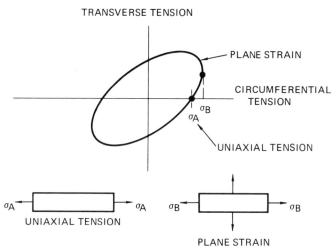

FIGURE 10-30

Schematic illustration of yield locus; the boundary between elastic and plastic deformation. Here σ_A represents the yield strength in tension; σ_B the yield strength in plane strain bending.

again on the yield locus. It is this increase in tensile stress needed to cause plasticity which reduces the ductility in bending or any plane strain process.

This can be thought of in terms of constant work to cause fracture. Work to cause fracture equals stress times strain. If we consider this work constant for a given material with a given processing history, the increasing tensile stress decreases the available ductility. We will now take a look at the materials variables that affect both stress and strain in bending in order to assess their effects on bend formability.

As noted earlier, the materials variables affecting bend formability are texture, ductility, and the ability to distribute strain. In the constant work to fracture concept, texture influences the stress term, and ductility and the ability to distribute stress influence the strain term.

Texture is described by a parameter termed R. We choose to represent texture by the R value because this parameter describes the effect of texture on the flow behavior of materials. R is the ratio of width strain to thickness strain measured at a given extension in a tensile test. An R value equal to 1 indicates the material is isotropic; there is no texture. Equal strains occur in both the width and the thickness direction. When

R values are different from 1, the strains in both directions are not equal; this indicates that a texture is present. Figure 10–31 shows this schematically. High *R* values encourage narrowing, while low *R* values encourage sheet thinning.

In bending, we want to encourage thinning and discourage narrowing in order to minimize the development of transverse tensile stresses, which impair ductility.

Figure 10–32 shows schematically the effect of changing *R* value on the shape of the yield locus. The curve for *R* = 1 represents the yield locus of an isotropic material (no texture present), and the curves for *R* greater than and less than 1 represent the yield loci of materials with texture. The *R* value can have a very strong effect on the shape of the yield locus and, therefore, on the strength in bending. The symbol σt represents the tensile strength (what the user buys) and σ_A, σ_B, and σ_C represent the strength in bending (what the user has to deal with in a bending application). In the constant work concept, low strength is desired to keep ductility as high as possible. Therefore metal with a low *R* value is desirable.

Figure 10–33 shows an example of how *R* value affects bend ductility. This figure is a plot of minimum bend radius versus ultimate tensile

TENSILE SPECIMEN BEFORE TEST

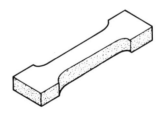

TENSILE SPECIMEN AFTER TEST

HIGH R R>1   LOW R R<1

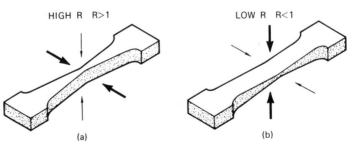

(a)   (b)

FIGURE 10–31
The effect of *R* on metal flow in a tensile test.

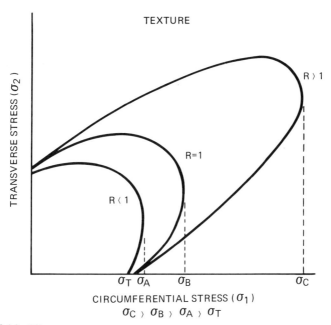

FIGURE 10–32

Schematic illustration of the effect of R value on the tension-tension quadrant of the yield locus. Here σ_T represents the yield strength in tension; σ_A, σ_B, and σ_C represent the yield strengths in plane strain bending for conditions of varying R value.

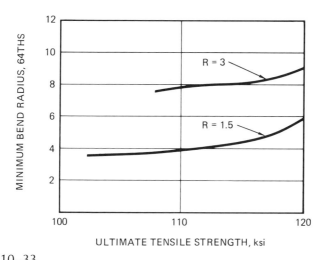

FIGURE 10–33

Minimum bend radius versus ultimate tensile strength curve for two different R values. The material is copper Alloy C63800, a cobalt modified aluminum-silicon bronze.

strength for copper Alloy C63800, a cobalt modified aluminum-silicon bronze. For a constant tensile strength, better bend formability is provided by material having a lower R value.

Figure 10–34 is a plot of minimum bend radius versus temper for Alloy C11000 and Alloy C51000. Bend curves for both good-way and bad-way directions are shown for each alloy. With increasing temper, the minimum bend radius increases for both alloys. However, Alloy C51000 develops marked bend anisotropy, while Alloy C11000 remains isotropic. This behavior is typical in that dilute alloys like C11000 tend to be isotropic, while solute strengthened alloys like C51000 tend to be anisotropic after cold work. This is due to the development of strong deformation textures in alloys like C51000. These alloys develop the (11) <112> brass type crystallographic texture. This texture has high R values in the bad-way direction which impair ducility. Therefore R values not only contribute to bend ductility but also are responsible for bend anisotropy.

Figure 10–35 shows how the ability to distribute strain, that is, the utilization of available ductility, can affect bend formability. This is a plot of outer fiber strain versus position along the bend surface for two materials. The critical strain to fracture in bending (E_c) is also shown. Material A is an ideal material with infinite capacity to distribute strain. In this case, every region of the bend surface reaches the critical strain level uniformly and the average strain is equal to the critical strain. Material B does not distribute strain uniformly. One element in the bend surface reaches the critical strain value and fractures sooner than the other elements. Accordingly, the average strain is lower than the critical strain. Therefore, for a given inherent ductility or critical strain to

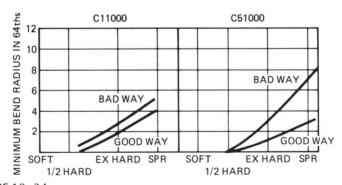

FIGURE 10–34
Good-way and bad-way minimum bend radius versus temper curves for Alloy C11000 copper and Alloy C51000 phosphor bronze.

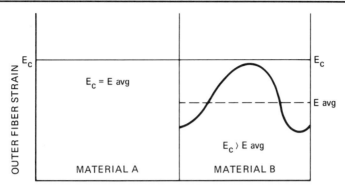

POSITION ALONG THE BEND SURFACE

FIGURE 10–35

Outer fiber strain versus position along the bend surface for ideal material A and real material B. Material A shows perfect strain distribution so that average strain (E_{avg}) equals critical strain to cause fracture (E_c). In real material B, the average strain is less than the critical strain.

failure, metal with a better capability to distribute strain will show better bend formability.

Figure 10–36 is a table which shows how critical strain and the ability to distribute strain affect bend formability for four different alloys. Alloy C72500 and Alloy C26000 have the same critical strain value and the same ability to distribute strain, and both show the same minimum bend radius. Alloy C11000 and Alloy C51000 show a lower minimum bend radius and better bend formability. Alloy C11000 achieves this by having a higher critical strain value; Alloy C51000, by using its ductility more efficiently.

ALLOY	CRITICAL STRAIN, %	BEND RADIUS IN 64ths	ABILITY TO DISTRIBUTE STRAIN
C72500	15	4	0.25
C26000	15	4	0.25
C11000	40	3	0.13
C51000	20	3	0.35

FIGURE 10–36

Influences of critical strain and strain distributing ability on minimum bend radius for four different copper alloys.

Figure 10–37 can be used to show how these three important factors—texture, ductility, and strain distributing ability—affect bend performance. This figure is a plot of minimum bend radius versus temper (percent cold reduction) for four alloys. Alloys C72500, C51000, and C63800 have increasing tendency to form deformation textures upon cold rolling. This texture development produces a high R value transverse to the rolling direction and is manifest as an increasing bend anisotropy from Alloy C72500 to C63800. Alloy C68800 and Alloy C63800 have approximately the same tendency to form deformation texture. However, Alloy C68800 has much better ability to distribute strain in the transverse direction than Alloy C63800. Accordingly, bad-way bend performance is improved so that anisotropy is not so marked.

Figure 10–38 illustrates alloy selection based on connector performance and fabrication criteria. The standard alloy selection method is used to compare minimum bend radius and yield strength. This is done in the lower curve of Figure 10–38 for Alloy C68800 and Alloy C51000.

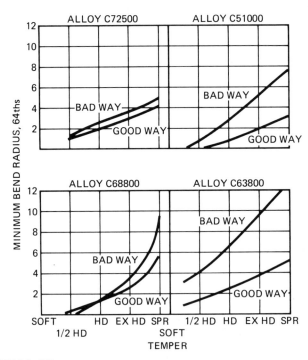

FIGURE 10–37
Minimum bend radius versus strength curves for four different copper alloys.

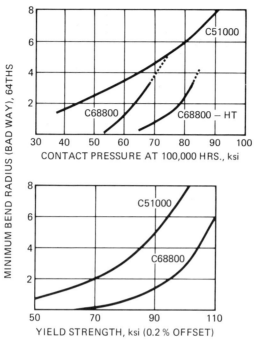

FIGURE 10–38
Alloy selection criteria—bottom curve shows minimum bend radius versus yield strength; top curve shows minimum bend radius versus contact pressure at 100,000 hours.

Alloy C68800 has better fabrication performance than Alloy C51000 at any given yield strength. This provides a tighter bend radius and improved design flexibility for miniaturization. For a given level of bend formability, Alloy C68800 has a higher yield strength, implying better connector reliability and performance. Accordingly, on the basis of bend versus yield strength, Alloy C68800 will be selected over Alloy C51000.

The upper curve of this figure illustrates what we believe is a more meaningful comparison of connector fabricability and performance for improved alloy selection. This plots minimum bend radius versus contact pressure at 100,000 hours instead of minimum bend radius versus yield strength. Contact pressure was derived from knowledge of the secant modulus and stress relaxation properties of these alloys, and the value at 100,000 hours was determined by extrapolation of short time (1000 hours) data. With this selection criterion, the comparison is somewhat changed in that Alloy C68800 provides better performance only up to contact pressures of 70 ksi; above 70 ksi, the two alloys are comparable. On the upper plot we have also indicated the performance of stabilized Alloy C68800. The stabilization heat treatment improves contact pressure significantly and provides the optimum bend formability-contact pressure combination.

Prediction of springback behavior is an important aspect of spring contact fabrication since part tolerance can affect the development of contact pressure. The amount of springback is determined by both geometry and materials variables. The geometry factors vary greatly from case to case and will not be discussed other than to say that solutions for most problems are available in the literature. Elastic properties are the most important materials factor determining springback. As Figure 10–39 shows, percent springback is equal to $1/E \times$ a function of geometry. Handbook values of elastic modulus E cannot be used with reliability. What is required is a secant modulus value determined on *unloading* from various total input deflections.

SPRING BACK

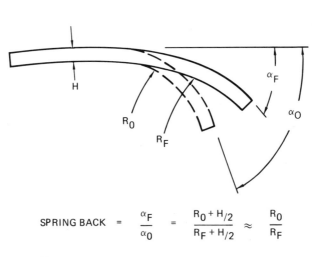

$$\text{SPRING BACK} = \frac{\alpha_F}{\alpha_0} = \frac{R_0 + H_{/2}}{R_F + H_{/2}} \approx \frac{R_0}{R_F}$$

$$\frac{R_0}{R_F} = 1 - \frac{1}{E} F(R_0, T)$$

FIGURE 10–39
Generalized equation for springback.

BIBLIOGRAPHY

1. American Society for Metals, *Metals Handbook,* 8th ed., Vol. 4: *Forming,* March 1970, p. 418.

2. American Society of Tool and Manufacturing Engineers, *Die Design Handbook,* McGraw-Hill Book Co., New York, 1955, Section 10.

3. Anaconda American Brass Company, "Metal drawing," Publication B-43.

4. Butts, A., *Copper,* Reinhold Publishing Corp., New York, 1954.

5. Crane, E. V., *Plastic Working in Presses,* 3rd ed. John Wiley & Sons, 1961, Chapters 8–10.

6. Doyle, L. E., C. A. Keyser, J. L. Leach, G. F. Schrader, and M. B. Singer, *Manufacturing Processes and Materials for Engineers,* 2nd ed., Prentice-Hall, Englewood Cliffs, N.J., 1969 Chapter 13.

7. Nordquist, Watson N., *Die Designing and Estimating,* 4th ed., Huebner Publications, Cleveland, Ohio, 1955, Chapters 10–12.

11

JOINING; WELDING, BRAZING, AND SOLDERING

Joining is a way of assembling copper alloys and copper alloy parts to provide useful articles which cannot be fabricated economically by other methods. Most of the known joining, bonding, and assembly techniques are applicable to many of the coppers and copper alloys. In this chapter some general information is presented on methods of joining, which should be a guide toward the solution of joining problems. Since this book is concerned with copper and copper alloy sheet and strip, emphasis is placed on metallurgical bonding techniques such as soldering methods, which are commonly used in joining these products.

As an initial aid in selecting a joining process the information contained in Table 11–1 is a guide for determining which methods are most applicable to copper and copper alloys. Table 11–2 suggests broadly how the design, joint configuration, and thickness of parts being joined should influence selection.

WELDING

In comparison to steels, copper alloys have high thermal conductivities and high coefficients of thermal expansion and contraction. Therefore welding may require preheating and high rates of heat input. Joint spacing is important and in general should be greater for copper alloys than for steel.

Oxyacetylene and *shielded metal-arc welding* once held primary positions for joining copper alloys. The gas shielded processes now have largely replaced them as these methods produce far superior welds. The

190

TABLE 11-1
Selection Guide for Joining Processes

Copper Alloy No.	Gas Metal-Arc	Gas Tungsten-Arc	Submerged-Arc	Shielded Metal-Arc	Stud	Electron Beam	Electroslag	Laser Beam	Gas	Resistance Spot	Resistance Seam	Projection	Flash	Brazing	Soldering	Adhesive Bonding	Mechanical Fastening
	Fusion Welding									Resistance Welding							
	Arc Welding					Other Welding Processes											
C10200	B	B	D	D	D	B/E	D	E	C	D	D	D	B	A	A	C	A
C11000, C11400	C	C	D	D	D	B/E	D	E	D	D	D	D	B	B	A	C	A
C21000, C22000	B	B	D	D	D	B/E	D		B	D	D	D	B	A	A	C	A
C22600, C23000	B	B	D	D	D		D		B	C	D	C	B	A	A	C	A
C26000, C26800	C	C	D	D	D		D		B	B	D	B	B	A	A	C	A
C44300	C	C	D	D	D		D		B	B	D	B	B	A	A	C	A
C31400, C35300	D	D	D	D	D		D		D	D	D	C	B	A	A	C	A
C36000	B	B	D	D	D		D		C	D	D	D	A	A	A	C	A
C54400	B	B	D	D	D		D		C	D	D	D	A	A	A	C	A
C51000, 52100	B	B	D	D	D		D		C	B	C	B	A	A	A	C	A
C61400	A	B	D	D	D		D		D	B	B	B	B	C	D	C	A
C65100	A	A	D	C	D		D		B	A	B	A	A	A	A	C	A
C65500	A	A	D	C	D		D		B	A	A	A	A	A	B	C	A
C67500	C	C	D	D	D		D		B	B	C	B	B	A	A	C	A
C68700	C	C	D	D	D		D		C	B	B	B	B	B	C	C	A
C70600	A	A	D	B	D		D		C	B	B	B	A	A	A	C	A
C71500	A	A	D	A	D		D		B	A	A	A	A	A	A	C	A
C75200, C77000	C	C	D	D	D		D		A	B	C	B	B	A	A	C	A

A. most satisfactory; B. satisfactory; C. restricted use; D. prohibited use; E. experimental.

tungsten inert-gas (TIG) and the *metal inert-gas* (MIG) are gas shielded processes, and metal thickness is the primary criterion determining the choice between the two. Typical operating data are shown in Tables 11–3 and 11–4.

Another widely used technique is *resistance welding* for thicknesses of 0.020 to 0.060 ins. Even though electrode sticking can occur and high power input is required, this method is popular because of the ease of adapting it to a production assembly line operation. Highly conductive copper alloys (>30% IACS) demand closely controlled operating procedures to produce good welds. As a comparative indication of resistance welding requirements, Table 11–5 lists typical copper alloys and their expected behavior.

BRAZING

Brazing is an effective means of joining copper alloys. Brazing filler metals with brazing temperatures of 625 to 1090°C (1150 to 2000°F)

TABLE 11-2a
Selection Guide for Joining Processes

	Fusion Welding									
	Arc Welding						Other Welding Processes			
Selection Criteria	Gas Metal-Arc	Gas Tungsten-Arc	Plasma-Arc	Submerged-Arc	Shielded Metal-Arc	Stud	Electron Beam	Electroslag	Laser Beam	Gas
A. Design Application										
Primary structural										
Elevated temperature	B	A	E	C	B	D	A	C	E	C
Ambient temperature	A	A	E	C	B	D	A	C	E	C
Cryogenic	B	A	E	D	C	D	A	D	E	D
Vacuum pressure	B	A	E	D	D	D	A	D	E	D
Atmospheric pressure	A	A	E	C	B	D	A	C	E	C
High pressure	B	A	E	D	D	D	A	D	E	D
Secondary structural	A	A	E	B	A	C	A	C	E	B
Noncritical	A	A	C	A	A	B	A	A	E	B
Dissimilar metal joining	C	C	C	D	B	C	C	D	E	C

A. most satisfactory; B. satisfactory; C. restricted use; D. prohibited use; E. experimental.

TABLE 11-2b
Selection Guide for Joining Processes

	Resistance Welding				Solid - State Welding							
Selection Criteria	Resistance Spot	Resistance Seam	Projection	Flash	Diffusion	Explosive	Ultrasonic	Friction	Brazing	Soldering	Adhesive Bonding	Mechanical Fastening
A. Design Application												
Primary structural												
Elevated temperature	A	A	D	A	A	C	C	D	B	D	C	B
Ambient temperature	A	A	D	A	A	C	C	D	A	C	A	A
Cryogenic	A	A	D	A	A	C	C	D	B	D	D	B
Vacuum pressure	D	B	D	C	A	C	C	D	B	D	C	C
Atmospheric pressure	B	A	D	B	A	C	C	D	A	D	C	C
High pressure	D	C	D	C	A	C	C	D	C	D	D	C
Secondary structural	A	A	C	A	A	C	C	C	A	D	A	A
Noncritical	A	A	A	A	A	C	B	B	A	B	A	A
Dissimilar metal joining	D	D	D	C	A	B	A	B	B	B	A	A

A. most satisfactory; B. satisfactory; C. restricted use; D. prohibited use; E. experimental.

TABLE 11-2c
Selection Guide for Joining Processes

| | Fusion Welding | | | | | | | | | |
| | Arc Welding | | | | | | Other Welding Processes | | | |
Selection Criteria	Gas Metal-Arc	Gas Tungsten-Arc	Plasma-Arc	Submerged-Arc	Shielded Metal-Arc	Stud	Electron Beam	Electroslag	Laser Beam	Gas
B. Joint Configuration										
Butt joint	A	A	A	A	A		A	A	A/E	B
Tee-joint	A	A	B	B	A		C	A/C	C/E	B
Edge joint	A	A	B	C	B		A	D	A/E	B
Corner joint	A	A	B	C	B		A	D	A/E	B
Flange joint	A	A	B	C	B		A	D	A/E	B
Scarf joint	D	D	D	D	D		C	D	D	D
Strap butt joint (splice joint)	C	C	C	C	C		C	D	B/E	D
Lap joint — Shear load	B	B	B	B	B		A	D	A/E	D
Lap joint — Tensile load	B	B	B	B	B		A	D	A/E	D

A. most satisfactory; B. satisfactory; C. restricted use; D. prohibited use; E. experimental.

TABLE 11-2d
Selection Guide for Joining Processes

| | Resistance Welding | | | | Solid - State Welding | | | | | | | |
Selection Criteria	Resistance Spot	Resistance Seam	Projection	Flash	Diffusion	Explosive	Ultrasonic	Friction	Brazing	Soldering	Adhesive Bonding	Mechanical Fastening
B. Joint Configuration												
Butt joint	D	D	D	A	C	D	D	B	D	D	D	D
Tee-joint	D	D	D	C	B	D	D	D	B	B	C	D
Edge joint	D	D	D	D	C	D	D	D	C	D	D	D
Corner joint	D	D	D	D	C	D	D	D	B	B	D	D
Flange joint	D	D	C	D	C	D	D	B	C	C	C	A
Scarf joint	D	D	D	D	C	B/E	D	D	B	B	B	B
Strap butt joint (splice joint)	B	B	C	D	C	C	B	D	A	B	A	A
Lap joint — Shear load	A	A	C	D	A	B	B	D	A	B	A	A
Lap joint — Tensile load	B	B	C	D	A	B	B	D	D	D	C	A

A. most satisfactory; B. satisfactory; C. restricted use; D. prohibited use; E. experimental.

TABLE 11-2e
Selection Guide for Joining Processes

Selection Criteria	Gas Metal-Arc	Gas Tungsten-Arc	Plasma-Arc	Submerged-Arc	Shielded Metal-Arc	Stud	Electron Beam	Electroslag	Laser Beam	Gas
C. Thickness of Parts Being Joined										
0.001 - 0.010 in.	D	B	A	D	D		A	D	B/E	D
0.010 - 0.020 in.	D	A	B	D	D		A	D	C/E	D
0.020 - 0.050 in.	C	A	B	D	D		A	D	D	B
0.050 - 0.100 in.	A	A	A	D	B		A	D	D	B
0.100 - 0.150 in.	A	A	A	D	A		A	D	D	B
0.150 - 0.250 in.	A	A	A	D	A		A	D	D	C
0.250 - 0.500 in.	A	A	C	C	A		A	D	D	D
0.500 - 1.000 in.	A	A	D	A	A		A	D	D	D
1.000 - 2.500 in.	B	B	D	A	A		A	A	D	D
2.500 in. over	C	B	D	A	A		A	A	D	D
Thick to thin	A	A	C	B	B		A	C	B/E	D

A. most satisfactory; B. satisfactory; C. restricted use; D. prohibited use; E. experimental.

TABLE 11-2f
Selection Guide for Joining Processes

Selection Criteria	Resistance Spot	Resistance Seam	Projection	Flash	Diffusion	Explosive	Ultrasonic	Friction	Brazing	Soldering	Adhesive Bonding	Mechanical Fastening
C. Thickness of Parts Being Joined												
0.001 - 0.010 in.	C	C	D	D	A	D	A	D	B	C	B	C
0.010 - 0.020 in.	A	A	B	D	A	D	A	D	B	B	A	B
0.020 - 0.050 in.	A	A	A	D	A	D	C	B	B	A	A	A
0.050 - 0.100 in.	A	A	A	A	A	D	D	B	B	C	A	A
0.100 - 0.150 in.	A	A	C	A	A	C	D	B	B	C	A	A
0.150 - 0.250 in.	C	C	C	A	A	C	D	B	C	D	B	A
0.250 - 0.500 in.	D	D	D	A	A	C	D	B	C	D	C	A
0.500 - 1.000 in.	D	D	D	A	A	C	D	B	D	D	D	A
1.000 - 2.500 in.	D	D	D	C	A	C	D	C	D	D	D	A
2.500 in. over	D	D	D	C	A	C	D	C	D	D	D	A
Thick to thin	C	C	C	C	A	A	A		B	A	A	A

A. most satisfactory; B. satisfactory; C. restricted use; D. prohibited use; E. experimental.

TABLE 11-3

Typical Operating Data for Tungsten Inert-Gas Welding

Metal Thickness, in.	Tungsten Electrode Diameter, in.	Helium Shield		Argon Shield		Preheat Temperature, °F
		Direct Current,[a] A	Gas, ft³/hr	Direct Current,[a] A	Gas, ft³/hr	
.020–.040	.030–.040	—	—	15–60	8–12	—
.040–.062	.040–.062	50–125	10–15	60–150	8–12	—
.125	.094	125–225	14–20	140–280	10–15	100
.188	.125	200–300	16–22	250–375	12–18	100
.250	.188	250–350	20–30	300–475	16–25	200
.500	.250	300–550	25–35	400–600	20–30	650
.750	.250	300–550	30–40	400–600	30–40	750
1	.250	300–600	30–40	450–650	30–40	750

[a]With alternating current use 60 to 70% of these values.
Data source—see Reference 2.

TABLE 11-4

Typical Operating Data for Metal Inert-Gas Welding

Metal Thickness, in.	Electrode Diameter, in.	Voltage, V	Current, A	Base Metal	Welding Rod
.062	.030	25–26	100–165	Cu	Deox. Cu or Cu-Si
.125	.035	26–27	100–200	Cu-Zn	Cu-Si
.188	.045	27–28	100–250	Cu-Si	Cu-Si
.250	.062	29–30	250–400	Cu-Sn	Cu-Si
.500	.078	31–32	300–450	Cu-Al	Cu-Al
.750	.094	33–34	400–500	Cu-Ni	Cu-Ni or Cu-Si
1	.188	34–35	400–500	Cu-Be	Cu-Be

Data source—see Reference 2.

have been used for joining copper alloys. Plate type transmission oil coolers, plate type ice maker heat exchangers, and copper alloy musical instruments are examples of brazed assemblies. Drawbacks of brazing are significant softening of the brazed metal, high cost of low temperature brazing filler metals, and the need for a controlled atmosphere furnace for low cost copper alloy brazing filler metals. Fluxing will almost always be required, and joint clearances of 0.001 to 0.005 in. should be used for the best joint strength and soundness. Table 11–6 contains some information on brazing coppers and copper alloys.

TABLE 11-5
Comparative Behavior of Various Alloys in Resistance Welding

Copper Alloy	Alloy Type	Electrical Conductivity, % IACS	Rating
C65100, C65500, C66100	Copper-silicon	7–12	Excellent
C70600, C71000, C71500	Copper-nickel	4–10	Excellent
C74500, C75200, C77000	Copper-nickel-zinc	5–9	Good
C60600, C61200, C61400, C62800, C63800, C68800	Copper-aluminum	7–18	Good
C52400	Copper-tin (D)	10–12	Fair
C66700	Manganese brass	15–16	Fair
C69400	Silicon red brass	15–16	Fair
C51000, C53200	Copper-tin	15–22	Fair
C26000	Cartridge brass	22–28	Fair
C27000	Yellow brass	22–28	Fair
C28000	Muntz metal	22–28	Fair
C46400	Naval brass	22–28	Fair
C67500	Manganese bronze	22–28	Fair
C24000	Low brass	32–43	Poor
C23000	Red brass	32–43	Poor
C22000	Commercial bronze	32–43	Poor
C19400, C19500	Modified coppers	50–65	Poor

Data source—see Reference 1.

SOLDERS

Solders used with copper alloys usually consist of tin and lead, with a 60% tin—40% lead mixture being the most common. The eutectic composition, 63% tin–37% lead, provides the lowest solder melting temperature at 183°C (361°F), well below that of either tin or lead. Other elements contained in the solder are either impurities that can be detrimental to the process, or elements added to improve the bond strength or alter some other characteristic in a beneficial manner. A representative listing of solder alloys is shown in Table 11–8 along with typical applications. Further information is contained in References 6 and 9.

Most copper alloys have good solderability. Surface characteristics and cleanness are the basic determinants of solderability. The metal should be free of dirt, oil, or other foreign matter. Metal oxides on the surfaces to be joined cause problems, requiring acid or alkaline cleaning treatments after annealing or extended storage to prepare the material for soldering. Fluxes clean and protect surfaces during soldering to improve wetting and expose the surfaces for bonding by the solder.

TABLE 11-6
Typical Operating Data for Brazing

Metal or Alloy	Type of Joint	Filler Metal	Flux	Temperature Range[a] °F, and Conditions	Procedure Notes
Copper					
Oxygen bearing	Lap (3x1)	BAg	As recommended for filler metal	1175–1400	Avoid reducing atmosphere.
Oxygen free	Lap	BCuP	None	1400–1700	Reducing atmosphere satisfactory.
Deoxidized		BCuAgP	None	1300–1500	
Special coppers					
Beryllium	Extremely clean	BCuP	Without high fluoride content	Lowest possible	Can be brazed and solution treated at 1450°F; fast-quench and age at 600–700°F.
Chromium	Lap	BCuP			Cannot heat-treat with brazing. Brazing must be combined with solution anneal.
Zirconium	Lap	BAg			
		BCuZn			
Copper-zinc (brasses)	Lap	BAg	AWS Type 3		Use silver filler metal on brasses with low copper content. For high copper content use silver or brass filler.
		BCuZn	AWS Type 5		
		BCuP	AWS Type 5		
Leaded brasses		BAg	Complete flux cover	Heat slowly	Difficult to braze when lead is over 3%. Stress-relief-anneal before brazing.
Copper-nickel-zinc (nickel silver)		BCuZn	Use plenty to prevent oxidation	Uniform temperature required	Stress-relieve before brazing.
Copper-tin (phosphor bronzes)	Lap	BAg	Needs adequate flux		Stress-relieve before brazing. Powder metal compacts require sealing pores before brazing.
	Lap	BCuP			
Copper-silicon (silicon bronzes)	Clean properly	BAg	Flux coat before brazing	Heat slowly and uniformly	Stress-relieve before brazing.
		BCuP			
Copper-nickel	Lap and butt	BCu	Flux thoroughly or furnace braze, red atmos.	2000–2100	Use BCuP filler metal on cupro-nickel higher than 30% with caution.
Copper-aluminum (aluminum bronzes)	Lap	BCuP	AWS Type 4		Alloys over 8% aluminum require special brazing fluxes and techniques.
		BAg			

[a]Exact range depends on alloy composition used.
Data source—see Reference 7.

TABLE 11-7
Soldering Compared to Other Joining Processes

Factor	Type of bond						
	Metallurgical			Mechanical			Chemical
	Solder	Braze	Weld	Crimp	Screw	Wrap	Conductive cement
Temp. limit of joint (melting or break-down). °C or (°F)	73-462 (100-800)	462-907 (800-1600)	Conductor melting temp.	No limit except that of wire			107-182 (160-300)
Heating effect on assembly	Small	Large	Small (quick)	None			Cures at ambient to (250) 157
Ease of rework and re-bonding of permanent joint	Simple		Not practical	Not practical	Simple		Not practical
Ease of joint relative to conductor	Small			Medium	Large	Small	Small
Process economy: Equipment cost	Low	Medium	High	Low			
Ease of automation	Easiest	More difficult	More difficult	More difficult			
Extra hardware?	No			Yes		No	No
Joint stable? Vibration	Yes			Yes	No	Yes	Yes
Oxidation	Yes			No		Yes	Yes

FLUXES

Flux reacts with the metallic oxides, dislodging and dissolving them, and enabling the clean metal surfaces to make solder contact. The type of flux selected should meet all the requirements of a particular alloy and application. There are three basic flux types: (1) *chloride or "acid" type,* (2) *organic type, and* (3) *rosin or resin type.* The first two are not used for soldering electrical connections. Both produce corrosive as well as hygroscopic residues and could damage an electrical circuit unless thoroughly washed. However, they are used extensively in applications where the assembly can be washed after soldering, as in automotive radiator assembly.

Rosin in its natural, solid state is noncorrosive but is a poor flux. It is activated by dissolution in organic solvents. The residue after soldering has an astronomical resistivity of 3300 trillion Ω/in.[3] and is not hygroscopic. Activated rosin fluxes are sometimes considered to be corrosive, but properly exposed to heat they are not. The use of these fluxes can definitely improve solderability and they can be used for

TABLE 11-8
Solders and Their Applications

Tin (Sn)	Lead (Pb)	Other Elements	Melting Point		ASTM B - 32 Spec.	Application
			°C	°F		
65	35		183-185	361-365		Electronic assemblies; dip and wave
60	40		183-188	361-370	60 A,B	soldering; lowest-melting tin-lead alloys
50	50		183-204	361-399	50 A,B	Hand-soldering radio and electrical equipment
50	48.5	1.5 Cu	183-215	361-419		Hand-soldering with unclad copper soldering bits to reduce bit erosion
30	70		183-255	361-491	30 A,B	Fuses, motors, cable jointing, pretinning
30	rem.	1-1.7 Sb	185-248	365-478	30C	baths
18	rem.	0.75-1 Sb	185-275	365-527		Electric lamp bases, motors, dipping baths
95		5 Sb	236-243	457-469	95 TA	For higher operating temperatures
5	93.5	1.5 Ag	296-301	565-574		than tin-lead alloys will withstand.
	97.5	2.5 Ag	305	581	2.5 S	The lead-rich alloys have inferior wetting
	94.5	5.5 Ag	304-365	579-689		properties

many applications where washing cannot be done. According to available information (Reference 9), "there has never been an authentic case of corrosion from the use of activated *rosin core* solder." Table 11–9 contains some data on fluxes.

SOLDERING

Soldering is one of the oldest methods of joining and is the most frequently used technique for coppers and copper alloys. It is particularly useful for joining electrical conductors. An indication of solder joint characteristics is given in Table 11–7, where three basic joining procedures for electrical connections are compared: metallurgical, mechanical, and chemical. Soldering is in the metallurgical category and has many advantages. It is useful and economical for joints where the service temperature is below 155°C (250°F). The solder acts as the bond between the parts and provides both electrical and mechanical continuity. To accomplish soldering, surface wetting and bonding must occur. These phenomena occur at the atomic or molecular level and are affected by surface contamination and material composition. Surface contamination will be discussed below, along with fluxes, solder, and soldering methods. The important aspects of the composition of a copper alloy with respect to solderability are given in the section on soldering methods.

TABLE 11-9a

Soldering Fluxes Listed in Order of Decreasing Chemical Activity

Classification and type	Typical fluxes	Vehicle	Use for these joints
Inorganic: Acids	Hydrochloric, hydro-fluoric, orthophosphoric	Water, petrolatum paste	Structural
Salts	Zinc chloride, ammonium chloride, tin chloride	Water, petrolatum paste, polyethylene glycol	Structural
Gases	Hydrogen, forming gas, dry HCL	None	Electrical
Organic (nonrosin base): Acids	Lactic, oleic, stearic, glu-tamic, phthalic	Water, organic solvents, petrolatum paste, poly-ethylene glycol	Structural, electrical
Halogens	Aniline hydrochloride, glu-tamic hydrochloride, bromide derivatives of palmitic acid, hydrazine hydrochloride (or hydro-bromide)	Same as organic acids	Structural, electrical
Amines and amides	Urea, ethylene diamine	Water, organic solvents, petrolatum paste, poly-ethylene glycol	Structural, electrical
Organic (rosin base): Activated	Water-white rosin, with activators	Isopropyl alcohol, or-ganic solvents, poly-ethylene glycol	Electrical
Water white	Water-white rosin only	Same as activated	Electrical

SURFACE PREPARATION

Surface cleanness plays a vital role in the joining of metals, and therefore, certain precautions or procedures are mandatory for good results. The removal of dirt, oil, grease, and oxides may be done by chemical or mechanical means or both. Solvent and alkali degreasing is generally recommended for the first three soils, and acid immersion for most copper alloy oxides. Possible mechanical methods include blasting, brushing, filing, scraping, and machining to obtain clean, uncontaminated surfaces. The choice of a method should be made carefully, keeping in mind the particular material and final application.

Another beneficial surface treatment sometimes used is plating with other metals to improve shelf-life solderability. This topic is discussed in the section entitled "Plating."

TABLE 11-9b
Soldering Flux Characteristics

Classification and type	Temperature stability	Tarnish removal	Corrosive-ness	Postsolder cleaning methods
Inorganic: Acids	Good	Very good	High	Hot-water rinse and neutralize; organic solvents; or degrease
Salts	Excellent	Very good	High	Hot-water rinse and neutralize; 2% HCL solution; hot-water rinse and neutralize; organic solvents; or degrease †
Gases	Excellent	Very good, at high temperatures	None normally	None required
Organic (nonrosin base): Acids	Fairly good	Fairly good	Moderate	Hot-water rinse and neutralize; organic solvents; or degrease †
Halogens	Fairly good	Fairly good	Moderate	Same as organic acids
Amines and amides	Fair	Fair	None normally	Hot-water rinse and neutralize; organic solvents; or degrease †
Organic (rosin base): Activated	Poor	Fair	None normally	Water-base detergents. Isopropyl alcohol: organic solvent; or degrease †
Water white	Poor	Poor	None	Same as activated w/w rosin but does not normally require post-cleaning

† For optimum cleaning, follow by demineralized and distilled water wash.

SOLDERING METHODS

There are a number of different ways to perform soldering, depending on the application. Basically, heat must be applied to both melt the solder and warm the surface to facilitate good wetting and flowing of the molten solder. Table 11–10 categorizes the most common methods with some of their advantages and disadvantages.

Solderability testing is often subjective. No standard, recognized, universal method has yet emerged. Two tests which are widely used are (1) a *spread test,* and (2) a *dip test.* In the spread test, a flat piece of material is first fluxed and heated to a prescribed temperature; then a prescribed amount of solder is placed on it, melted, and allowed to spread; after it has cooled, the solder covered area is measured. In the dip test, the sample material is immersed in flux and then in a solder bath, from

TABLE 11-10
Comparison of Various Soldering Methods

Method	Advantages	Disadvantages
1. Soldering iron	Easily applied, inexpensive	Slow, only small parts
2. Dip and wave	High volume rates with many connections simultaneously	Solder contamination, critical adjustment
3. Flame	For large masses, portability	Open fire, overheating, little control of solder flow
4. Induction heating	For large masses, localized heat control, high volume, and quality joints	Needs extreme cleanness, good part clearances
5. Electrical resistance	Localized heating, useful when unsuited for soldering irons	Small volume, small assemblies
6. Oven heating	For large, complicated assemblies and mass production	Expensive to set up
7. Ultrasonic	Removes surface oxides and thus eliminates fluxes	Only for small areas; will not solder lap or crimp joints

which it is slowly withdrawn. The coverage and appearance are visually rated on a good to bad basis and classified (Classes I to V). The spread test eliminates operator judgment and correlates with direct-heat soldering methods (e.g. a soldering iron), whereas the dip test correlates with dip- or wave-soldering techniques.

Although clean copper rates as the easiest of all metals to solder, its solderability, along with that of the copper alloys, degrades with time (so-called *shelf-life solderability*) because of gradual surface oxidation and tarnishing. All of the copper alloys tend to tarnish when exposed, so none exhibits outstanding performance in the bare condition after storage, without the use of a flux that will remove the tarnish. (See earlier section on fluxes.) Plating the metal with a good tarnish-resistant metal (e.g. gold, tin, or solder) enhances shelf-life solderability. Copper-zinc alloy solderability loss during storage varies proportionately with zinc content, degrading more rapidly as the percentage of zinc increases. The fact that zinc exhibits high diffusion mobility and high oxidation potential explains this effect on solderability. It can even diffuse through and oxidize on the surface of plating materials, used on such alloys, when plating is thin and storage time is long. Although copper-zinc alloys present soldering problems after long storage, copper also oxidizes, as do other elements like tin and aluminum. As a guide, Table 11–11 describes the relative behavior of some copper alloy families, and notes the effects of the alloying elements on solderability.

TABLE 11-11
Relative Solderability of Various Alloy Types

Metal or Alloy	Solderability and Remarks
Coppers (Including tough pitch oxygen-free phosphorized, arsenical, silver-bearing leaded, tellurium, and selenium copper)	Excellent. Need only rosin or other noncorrosive flux.
Copper-tin	Good. Easily soldered with activated rosin and intermediate fluxes.
Copper-zinc	Good. Easily soldered with activated rosin and intermediate fluxes.
Copper-nickel	Good. Easily soldered with intermediate and corrosive type fluxes.
Copper-chromium and beryllium-copper	Good. Require intermediate and corrosive type fluxes.
Copper-silicon	Fair. Silicon produces refractory oxides that require use of corrosive fluxes.
Copper-aluminum	Difficult. May be soldered with help of very corrosive fluxes.
High-tensile manganese bronze	Not recommended. Should be plated to insure consistent solderability.

Data source—see Reference 9.

In order to provide information comparing the shelf-life solderability of various alloys, a set of graphs follows, Figures 11–1 through 11–3), along with explanatory Table 11–12. These figures summarize the results of tests performed. (See Reference 10.) The tin plated specimens were electrotinned with approximately 200 μin. or 5 microns of tin, which is the typical thickness produced by such a process. The tin definitely improves shelf life, but degradation of the solderability of the zinc-bearing alloys eventually occurs because of zinc migration and consequent oxide formation. The initial solderability of these alloys when tin coated, is no better than when in their bare surface condition, because the tin melts during soldering, mixing with the solder as it covers the base metal.

PLATING

Stringent shelf-life solderability requirements demand some type of plating to insure good performance after long time periods and expo-

TABLE 11-12
Solderability Rating System Used in Figures 11-1, 11-2, and 11-3

Class I
An "ideal" coating. The solder layer is smooth and uniformly thick with no surface irregularities or gaps.

Class II
Also an essentially continuous solder coating without solder dewetting or solder pullback. Less than 1% of the surface may be taken up by "pinholes" (small areas where the solder did not wet the surface). The solder coating is rougher than a Class I coating; it need not be of equal thickness at all points on the specimen.

Class III
Evidence of solder pullback and/or dewetting on up to 50% of the specimen surface. Up to 10% of the surface may be bare.

Class IV
Dewetting and/or pullback over 50% or more of the surface. More than 10% of the surface is bare.

Class V
No solder adhesion to the surface except for small amounts, usually in the form of droplets, which can be readily removed mechanically.

sure to different environments. A noble metal like gold provides adequate protection but at a high economic penalty. With this limitation in mind, considerable experimentation has been initiated employing tin and solder plating, possibly with a copper or nickel undercoat, to improve shelf-life solderability. Many data have been collected, and some results compiled by the Bell Laboratories were selected to illustrate this subject. Tables 11–12 through 11–15 indicate the solderability performances of three representative copper alloys: C11000, ETP copper, C26000, cartridge brass, and C76200, nickel silver. The plated specimens were exposed to three different environments: (1) severe industrial—factory location, (2) mild industrial—air-conditioned factory location, and (3) accelerated aging—exposure to 38°C (100°F) at 85% relative humidity. After exposure, a dip test using unactivated flux was performed. Ratings for solderability of A to D correspond to the class values in Table 11–12, that is, A = I, B = II, C = III, and D = IV and V; classes C and D are considered unsuitable. Overall numerical ratings were calculated by totaling each row, setting A = 10, B = 8, and C or D = 0.

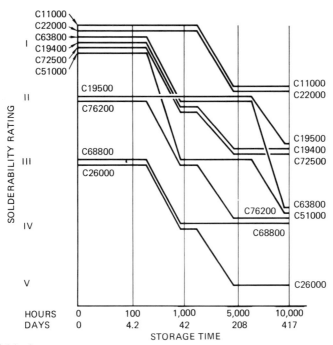

FIGURE 11–1
Solderability ratings of bare copper alloy strip, using a mildly activated flux. (Zarlingo [8])

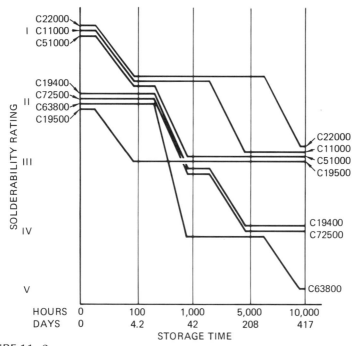

FIGURE 11–2
Solderability ratings of bare copper alloy strip, using a nonactivated flux. (Zarlingo [8])

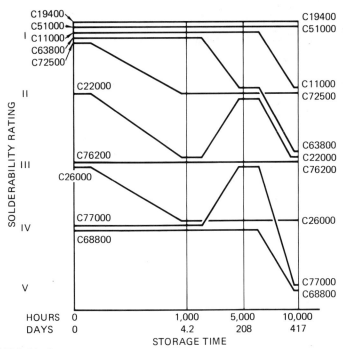

FIGURE 11-3

Solderability ratings of tin-plated copper alloy strip, using a nonactivated flux. (Zarlingo [8])

TABLE 11-13

Solderability Ratings Before and After Exposure

| | Thickness in μm | | | | Control | Exposure Site and Time of Exposure in Months | | | | | | | | | |
| | Undercoat | | Final Coat | | | Severe Industrial | | | Mild Industrial | | | Aging Test | | | |
	Copper	Nickel	Tin	Solder	0	3	6	12[a]	3	6	12[a]	3	6	12[a]	Rating
C11000	-	-	0.6	-	A	A	B	D	A	A	D	D	D	D	48
	-	-	-	0.6	A	C/D	C/D	D	A	B	C	D	D	D	28
	-	-	-	1.3	A	A	A	A	A	A	A	A	A	A	90
	-	-	-	2.5	A	A	A	A	A	A	A	A	A	A	100
	-	-	-	5.0	A	A	A	A	A	A	A	A	A	A	100
C26000	-	-	2.5	-	A	D	C/D	D	B	C	D	D	D	D	18
	-	-	-	2.5	A	A	A	A	A	A	A	A	A	B/C	90
	2.5	-	-	2.5	A	A	A	A	A	A	A	A	A	A	100
	2.5	-	5.0	-	A	A	A	A	A	A	A	A	A	B	98
	5.0	-	1.3	-	A	A	C	D	A	A	C	A	D	D	50
	5.0	-	2.5	-	A	A	A	B	A	A	A	A	A	C	88
	5.0	-	5.0	-	A	A	A	A	A	A	A	A	A	B	98
	-	1.3	2.5	-	A	A	A	B	A	A	A	A	B	D	86
	-	2.5	1.3	-	A	A	(B)[b]	B	A	A	A	A	A	C	86
	-	2.5	5.0	-	A	A	A	A	A	A	A	A	A	B	98
	-	2.5	-	2.5	A	A	A	A	A	A	A	A	A	A	100
	-	2.5	-	5.0	A	A	A	A	A	A	A	A	A	A	100
	-	2.5	2.5	-	A	A	A	C	A	A	A	A	(B)	(C)	78

[a]Includes 12 months exposure at site plus additional 12 months storage in a temperature and humidity controlled laboratory at authors laboratory before solder dip test evaluation. All panels wrapped and protected during Murray Hill storage.

[b]Letters in parentheses indicate reasonable rating for a missing panel.

[c]Tin plate was reflowed in an oil bath at a temperature of 250C (500F) before exposure.

206

TABLE 11-13 (cont.)

| | Thickness in μm | | | | Exposure Site and Time of Exposure in Months | | | | | | | | | |
| | Undercoat | | Final Coat | | Control | Severe Industrial | | | Mild Industrial | | | Aging Test | | | |
	Copper	Nickel	Tin	Solder	0	3	6	12[a]	3	6	12[a]	3	6	12[a]	Rating
C26000	1.3	-	2.5	-	A	A	A	A	A	A	A	A	A	B/C	90
	2.5	-	1.3	-	A	C	C	D	A	A	B	A/B	C	D	46
(Tin-Oil Reflowed)[c]	2.5	-	2.5	-	A	A	B	B	A	A	A	A	A	B/C	86
C76200	1.3	-	2.5	-	A	A	B	B/C	A	A	A	A	B	D	76
	2.5	-	1.3	-	A	A	C	D	A	A	A/B	A/B	C	D	56
	2.5	-	2.5	-	A	A	B	D	A	A	A	A	B	D	76
	2.5	-	5.0	-	A	A	A	A	A	A	A	A	A	B	98
	-	-	-	0.6	C	C	C	C	C	C	C	C	C	D	0
	-	-	-	2.6	B	C	C	C	B/C	B/C	B/C	B	B	C	24

[a]Includes 12 months exposure at site plus additional 12 months storage in a temperature and humidity controlled laboratory at authors laboratory before solder dip test evaluation. All panels wrapped and protected during Murray Hill storage.

[b]Letters in parentheses indicate reasonable rating for a missing panel.

[c]Tin plate was reflowed in an oil bath at a temperature of 250C (500F) before exposure.

TABLE 11-14
Grouping of Solderability Ratings for the Three Exposures of Table 11-13

| Group Solderability Rating | Basis Metal | Thickness in μm | | | | Numerical Solderability Number |
| | | Undercoating | | Final Coating | | |
		Copper	Nickel	Tin	Solder	
100	Copper	-	-	-	2.5	100
	Copper	-	-	-	5.0	100
	Brass	2.5	-	-	2.5	100
	Brass	-	2.5	-	2.5	100
	Brass	-	2.5	-	5.0	100
90-99	Brass	2.5	-	5.0	-	98
	Brass	5.0	-	5.0	-	98
	Brass	-	2.5	5.0	-	98
	Nickel Brass	2.5	-	5.0	-	98
	Steel	7.5	-	10.0	-	98
	Copper	-	-	-	1.3	90
	Brass	-	-	-	2.5	90
	Brass [a]	1.3	-	2.5 [a]	-	90 [a]
80-89	Brass	5.0	-	2.5	-	88
	Steel	5.0	-	5.0	-	88
	Brass	-	1.3	2.5	-	86
	Brass [a]	2.5	-	2.5 [a]	-	86 [a]
	Brass	-	2.5	1.3	-	86

[a] The tin plate was reflowed in oil bath at a temperature of 250°C (500°F) before exposure.

TABLE 11-14 (cont.)

Group Solderability Rating	Basis Metal	Undercoating Copper	Undercoating Nickel	Final Coating Tin	Final Coating Solder	Numerical Solderability Number
70-79	Brass	-	2.5	2.5	-	78
	Nickel Brass	1.3	-	2.5	-	76
	Nickel Brass	2.5	-	2.5	-	76
	Steel	2.5	-	2.5	-	74
60-69	Steel	-	-	2.5	-	63
50-59	Nickel Brass	2.5	-	1.3	-	56
	Steel	5.0	-	1.3	-	56
	Brass	5.0	-	1.3	-	50
40-49	Copper	-	-	0.6	-	48
	Brass [a]	2.5	-	1.3 [a]	-	46 [a]
30-39	NONE					
20-29	Copper	-	-	-	0.6	28
	Nickel Brass	-	-	-	2.5	24
10-19	Brass	-	-	2.5	-	18
0-10	Nickel Brass	-	-	-	0.6	0

[a] The tin plate was reflowed in oil bath at a temperature of 250°C (500°F) before exposure.

TABLE 11-15
Solderability Ratings for Severe Industrial and Mild Industrial Exposures

Solderability Ratings in Groups	Basis Metal	Undercoating Copper	Undercoating Nickel	Final Coating Tin	Final Coating Solder	Actual Ratings Obtained
Group 100	Copper	-	-	-	1.3	100
	Copper	-	-	-	2.5	100
	Copper	-	-	-	5.0	100
	Brass [a]	1.3	-	2.5 [a]	-	100 [a]
	Brass	-	-	-	2.5	100
	Brass	2.5	-	-	2.5	100
	Brass	2.5	-	5.0	-	100
	Brass	5.0	-	5.0	-	100
	Brass	-	2.5	-	2.5	100
	Brass	-	2.5	-	5.0	100
	Brass	-	2.5	5.0	-	100
	Nickel Brass	2.5	-	5.0	-	100
	Steel	7.5	-	10.0	-	100
Group 90-99	Brass	5.0	-	2.5	-	97
	Brass	-	1.3	2.5	-	97
	Steel	5.0	-	5.0	-	97
	Brass	-	2.5	1.3	-	94
	Brass [a]	2.5	-	2.5 [a]	-	94 [a]

[a] The tin plate was reflowed in oil bath at a temperature of 250°C (500°F) before exposure.

TABLE 11-15 (cont.)

| Solderability Ratings in Groups | Basis Metal | Thickness in μm | | | | Actual Ratings Obtained |
| | | Undercoating | | Final Coating | | |
		Copper	Nickel	Tin	Solder	
Group 80-89	Brass	-	2.5	2.5	-	88
	Nickel Brass	1.3	-	2.5	-	83
	Nickel Brass	2.5	-	2.5	-	83
	Steel	2.5	-	2.5	-	80
Group 70-79	Steel	-	-	2.5	-	71
Group 60-69	Copper	-	-	0.6	-	69
	Nickel Brass	2.3	-	1.3	-	69
	Steel	5.0	-	1.3	-	68
Group 50-59	Brass	5.0	-	1.3	-	57
	Brass [a]	2.5	-	1.3 [a]	-	54 [a]
Group 40-49	Copper	-	-	-	0.6	40
Group 30-39	NONE					
Group 20-29	Brass	-	-	2.5	-	26
Group 10-19	Nickel Brass	-	-	-	2.5	11
Group 0-9	Nickel Brass	-	-	-	0.6	0

[a] The tin plate was reflowed in oil bath at a temperature of 250°C (500°F) before exposure.

The general conclusions drawn from these tests were (1) zinc-bearing alloys require at least 2.5 microns of copper or nickel under a minimum of 2.5 microns of tin or 1.3 microns of solder; (2) solder appears to be better than tin coating; (3) some type of plating must be incorporated to guarantee any chance of satisfactory solderability after long storage time or exposure to hostile environments; and (4) use of an activated rosin flux can improve the solderability in many cases.

Other data on this subject are available, but in general they duplicate most of the results included here. Research is continuing in this field to provide further information for material selection. Hopefully, a new technique, a new solder formulation, a guaranteed noncorrosive flux, or some other breakthrough will provide satisfactory solderability for all copper alloys. Until then, each alloy and application must be programmed using existing technology to obtain satisfactory soldering results.

BIBLIOGRAPHY

1. American Welding Society, *Welding Handbook,* 1972.
2. "How to Weld Copper," *Steel,* January-March 1958.
3. Koehler, M. L., et al., "Guidelines for Joining Process Selection," *Met. Prog.,* November 1972, pp. 63, 64.
4. Manko, H. H., "How to Choose the Right Solder Flux," *Prod. Eng.,* June 13, 1964.
5. Manko, H. H. "How to Design the Soldered Electrical Connection," *Prod. Eng,* June 12, 1972.
6. Manko, Howard H., *Solders and Soldering,* McGraw-Hill Book Co., New York, 1964.
7. "On Joining Copper," *Welding Desi Fabri,* February 1973.
8. Zarlingo, S. P., "Materials for Electrical and Electronic Applications (Solderability and Wear Resistance)," *Proceedings of the Nineteenth Annual Holm Seminar on Electric Contact Phenomena,* 1973.
9. Kester Solder Company, *"Solder,"* 1961.
10. "Solder and Tin Coatings After Extended Storage," *Plating,* March, 1973.
11. American Welding Society, *Soldering Manual,* 1959.

12

SURFACE TREATMENTS: COLOR AND COLORING, POLISHING, BUFFING, PATTERNED METALS

Apart from their durability, adaptability to an endless variety of uses, and ease of fabrication, their natural range of warm colors have given copper and its alloys artistic appeal and aesthetic favor for centuries. This chapter reviews some of the compositional factors affecting the inherent color of these metals, methods whereby they may be chemically colored, and procedures for polishing and buffing them. The importance of choosing the correct surface finishes for various end uses is discussed. An embossed surface is a special rolling mill finish that is included along with the various finishes applied to parts fabricated from strip.

COLOR

Other than gold, copper is the only metal that is not naturally almost white or gray in color. Architects and designers specify copper alloys because of their beautiful natural colors and low cost. Years of experience have resulted in the development of improved techniques for treating copper alloy surfaces. They have made available a wide variety of hues in reds, yellows, greens, grays, browns, blacks, and combinations thereof.

Besides unalloyed copper, which is naturally red, the primary copper alloy families whose colors are red or gold are the bronzes (copper-tin) and the brasses (copper-zinc). The *chroma* of these alloys, produced as strip, derives from the red hue of the copper. Both tin and zinc affect color tones as a function of their concentrations. Although true bronzes are alloys of copper and tin, the low zinc-high copper brasses have bronze colors. These alloys have common names which are indicative of their colors and some of their applications. Alloy C21000 (95 Cu-5 Zn) is commonly called Gilding Metal. It has frequently been used for severely coined medallions and emblems, which are often gold plated or vitreous enameled. Alloy C22000 (90 Cu-10 Zn), called Commercial Bronze, has a distinctive bronze color and is used as weatherstrip and lock and door hardware because its color is close to that of architectural bronze. Alloy C22600 (87 Cu-13 Zn) is called Jewelry Bronze and has a color which is similar to that of gold. It finds wide application in costume jewelry and as a base for gold plated jewelry. Alloy C23000 (85 Cu-14 Zn) is similar to gold in color although its common name is red brass. It is used for jewelry and many other decorative applications where its beautiful gold color is a basic requirement. Among such applications are gold colored picture frames.

In wrought copper alloy sheet and strip, the copper-zinc alloys from 5 through 30% zinc provide this range of warm colors at a moderate cost and so are frequently chosen for applications on the basis of color. Since the tin bronze alloys containing up to 8% tin are considerably more costly, they are less likely to be chosen for color alone. However, a pen manufacturer, known for the excellent quality of his product, uses an 8% tin phosphor bronze strip to fabricate beautiful gold-colored pen clips. Here, the excellent formability and the strength of the alloy are necessary properties, as well as the gold color.

The nickel silvers are often used for decorative as well as utilitarian applications. Their silver colors are tinged with hues which range from pink to yellow to green, blue, and silver white, depending upon the percentages of zinc and nickel which are alloyed with the copper. Figure 12–1 illustrates the relationship between composition and color of the nickel silvers. Silver plated table flatware and hollow ware are usually made from nickel silvers. The alloys used are capable of being severely worked and have high strength and a silvery color. With the silver colored base metal under the silver plate, wear and incidental handling damage during long years of service detract very little from the aesthetic value. Silver plated musical instruments also frequently are nickel silver under the silver plating. Optical-frame hardware and woodwind-instrument keys are other applications. The ease with which the color can be maintained on such parts is another reason for the choice of nickel silver.

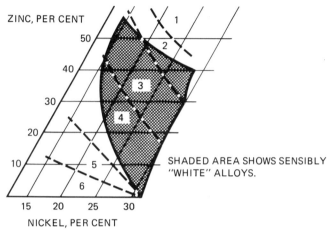

1 DEFINITE BLUE CAST
2 WHITE WITH SLIGHT BLUISH CAST
3 WHITE WITH PALE GREEN CAST
4 DEFINITE TO SLIGHT GREEN CAST
5 DEFINITE TO SLIGHT YELLOW CAST
6 DEFINITE TO SLIGHT PINK CAST

ZINC, PER CENT

SHADED AREA SHOWS SENSIBLY "WHITE" ALLOYS.

NICKEL, PER CENT

FIGURE 12–1

Relationship between composition and color of nickel silver alloys (Kihlgren et al. [2])

Although the copper alloys are resistant to corrosion by normal atmospheric conditions, they do become tarnished and discolored unless protected. Articles such as brass and bronze door hardware are lacquered with colorless transparent lacquers, baked on and specially formulated to resist wear and maintain color for long periods of time. Since this discussion could not cover all the different protective coverings which might be formulated for the many applications of copper alloys, the interested reader is referred to the manufacturers of such products and to the literature on the finishing of metals for more specific information. The Copper Development Association has sponsored a lengthy study of clear coatings, and reports on it can be obtained from this organization.

COLORING BY CHEMICAL TREATMENT

Copper has been used as a roof covering for centuries, and bronze has been used for statuary. As they weather, a beautiful green or blue-green *patina* develops on them. The development of artificial means of hastening the weathering process has been pursued, and information is available from the Copper Development Association. Other processes for coloring copper and copper alloys to provide the appearance of grained wood, or an antique appearance, or various other colors have also been used by the architect and artisan. Such finishes and methods can be found in platers' and finishers' handbooks and guides. A few examples of such treatments, taken from Reference (2) at the end of this chapter, are given below.

1. *Verde Antique on Copper or Brass*

   The following is a method to produce a patina or verde antique effect. The green bronze antique tones are the result of a combination of chemicals, pigments, and artistic skill. The finish is quickly obtained and may be used on brass or copper. It can be stippled onto a plain surface, or dipped on background work where the high parts are to be relieved, and is adapated to large surfaces. The green will work out over the metal.

   The solution is as follows:

Nitrate of copper	4 oz
Sal ammoniac	4 oz
Calcium chloride	4 oz
Water chloride	1 gal

   In stippling the solution onto large surfaces, to prevent runs a paste can be made by mixing a little carbonate of copper with this liquid. To stipple, it is best to use a stiff bristle brush, set in rubber or plastic, as the iron binding on an ordinary brush is attacked by the acid in the solution and causes spots (or rust) in the finish.

   The green will appear in a short time and should be lacquered afterward for protection. This will give the green a dull finish and will prevent a gloss which often gives to verde antique finish the appearance of green paint. If a waxed effect is desired, the waxing can be done either with beeswax over lacquer, using it on a tampico wheel brush run at slow speed, or with paraffin wax cut in Venice turpentine to a liquid, applying it with a brush.

2. *Antique Green Oxidized*

   After cleaning the following solution should be used:

Water	1 gal
Iron chloride	3 oz
Sal ammoniac	16 oz
Verdigris powder	8 oz
Common salt	10 oz
Cream of tartar	4 oz

The solution should be applied with a brush or dip and then stippled with a soft, round brush to give the variegated appearance of naturally aged bronze.

3. *Hardware Green on Brass*

   A fine emery finish should be given first. Then the article should be cleaned thoroughly and immersed in the following solution until the brass develops a greenish color:

Water (180°F)	1 gal
Thiosulphate of soda ("Hypo")	8 oz
Nitrate of iron	2 oz

   When a green tone develops, the article should be washed in water and highlights touched up, using a tampico brush or wheel and a little fine brimstone and water.

4. *Brown on Yellow Brass*

   Surface: Buffed.

   Preparation: Hot alkaline cleaner, cold water rinse; cyanide dip, cold water rinse.

   Solution used;

Sodium bichromate	150 g
Nitric acid (sp. gr. 1.42)	20 cc
Hydrochloric acid (sp. gr. 1.20)	5 cc
Aerosol AY	0.75 g
Water	1000 cc

   Procedure: Immerse 1 minute at room temperature (70° F) with agitation of piece at 15 second intervals. Rinse in cold and hot water; dry by air blast. Lacquer with a clear nitrocellulose resin.

   Remarks: This film is easily removed when wet; after drying it is very adherent if properly applied.

5. *Gold on Yellow Brass*

   Surface: Buffed.

   Preparation: Hot alkaline cleaner, cold water rinse; cyanide dip, cold water rinse.

   Solution used:

Sodium bichromate	150 g
Nitric acid (sp. gr. 1.42)	20 cc
Hydrochloric acid (sp. gr. 1.20)	5 cc
Sulfuric acid (sp. gr. 1.84)	3 cc
Aerosol AY	0.75 g
Water	1000 cc

Procedure: Immerse 1 minute at room temperature (70° F.) with agitation of piece at 15 second intervals. Rinse in cold and hot water; dry by air blast. Lacquer with a clear nitrocellulose resin.

Remarks: This film is easily removed when wet; after drying it is very adherent when properly applied.

6. *Light Green on Yellow Brass*
   Surface: Bright-dipped.
   Preparation: Hot alkaline cleaner, cold water rinse; cyanide dip, cold water rinse.
   Solution used:

Sodium bichromate	150 g
Phosphoric acid (sp. gr. 1.71)	10 cc
Aerosol AY	0.75 g
Water	1000 cc

   Procedure: Immerse 10 to 15 minutes with occasional agitation of piece. Rinse in hot and cold water; dry by air blast. Lacquer with a clear nitrocellulose resin.

   Remarks: This film is easily removed when wet; after drying it is very adherent if properly applied.

7. *Blue-Black on Yellow Brass*
   Surface: Buffed or bright-dipped.
   Preparation: Hot alkaline cleaner, cold water rinse; cyanide dip, cold water rinse.
   Solution used:

Copper carbonate	120 g
Ammonium hydroxide (sp. gr. 0.90)	250 cc
Water	750 cc

   An excess of copper carbonate must be present.
   Procedure: Immerse approximately 10 seconds at 180 to 200° F. Follow with cold water rinse, alkaline rinse, cold water rinse, alcohol rinse; dry in clean sawdust. Lacquer with a clear nitrocellulose resin.

8. *Steel Gray on Yellow Brass*
   Surface: Bright-dipped.
   Preparation: Hot alkaline cleaner, cold water rinse; cyanide dip, cold water rinse.
   Solution used:

Arsenic trioxide (As_2O_3)	30 g
Hydrochloric acid (sp. gr. 1.20)	65 cc

Sulfuric acid (sp. gr. 1.84)   16 cc
Water   1000 cc

Procedure: Immerse 5 to 10 seconds in solution at room temperature (70°F.). Rinse in cold and hot water; dry by air
blast. Lacquer with a clear nitrocellulose resin.

9. *Statuary Bronze on Yellow Brass*
   Surface: Bright-dipped.
   Preparation: Hot alkaline cleaner, cold water rinse; cyanide dip,
   cold water rinse.
   Solution used:

Copper carbonate   120 g
Ammonium hydroxide (sp. gr. 0.90)   250 cc
Water   750 cc

An excess of copper carbonate must be present.
Procedure: Immerse approximately 10 seconds at 180 to 200°F.
Rinse in cold water. Develop a brown color by immersion for a few seconds in a dilute solution of sulfuric
acid (15 cc). Rinse thoroughly in cold water; dry by air
blast. Remove smut with a clean, soft rag or sawdust.
Lacquer with a clear nitrocellulose resin.

10. *Black on Yellow Brass*
    Surface: Bright-dipped.
    Preparation: Hot alkaline cleaner, cold water rinse; cyanide dip,
    cold water rinse.
    Solution used
    and Procedure: Suspend the piece in a hot caustic soda solution (60
    g/li) for a few minutes; then transfer to another
    caustic soda solution of the same strength to which
    7.5 g potassium persulfate/li has been added and
    heated to incipient boiling. Immerse the piece for
    10 minutes, remove, and rinse thoroughly in cold
    water and in hot water; then air-dry. Smooth the
    black-velvet-like film by rubbing with a clean, soft
    cloth. Lacquer with a clear nitrocellulose resin.

11. *Black on Copper*
    Surface: Buffed.
    Preparation: Hot alkaline cleaner, cold water rinse; cyanide dip,
    cold water rinse.
    Solution used:

Potassium sulfide   15 g
Water   1000 cc

        Procedure: Immerse for 5 to 10 seconds in the above solution at 100°F. Rinse in cold and hot water; dry by air blast. Lacquer with a clear nitrocellulose resin.

12. *Royal Copper on Copper*
    Surface: Bright-dipped. Buffed.
    Preparation: Hot alkaline cleaner, cold water rinse; cyanide dip, cold water rinse; dry by air blast.
    Procedure: Immerse in molten potassium nitrate, in an iron container, at 1200 to 1300°F (Dark Red) for 20 seconds. Remove and quench in hot water, dry, and buff. Lacquer with a clear nitrocellulose resin.

13. *Blue on Brass*
    Solution used:

White arsenic	16 oz
Water	0.5 gal
Muriatic acid	1 gal

    Procedure: Immerse metal into the solution until the desired light blue finish results.

A most important prerequisite is that the surface to be colored be clean—free of dirt, oil, grease, and oxides. The oily contaminants can be removed with a good chemical cleaner of the trisodium phosphate type. Any oxides on the surface may require cleaning in a dilute sulfuric acid solution. A 10% sulfuric acid solution warmed to around 126°F is a common "pickling" agent. Stronger acids such as the bright dip described below may be used for more stubborn, heavier oxides.

As a possible finishing touch, a bright dip, consisting of nitric, sulfuric, and small amounts of hydrochloric acid, can be used to impart a brighter, passivated surface that will readily accept color treatments. A typical formula is:

Sulfuric acid (sp.gr. 1.83)	50–60% by volume
Nitric acid (sp.gr. 1.41)	15–25% by volume
Hydrochloric acid (sp.gr. 1.16)	0.5 oz/gal
Water	Remainder

Use at room temperature; immerse for 5 to 45 seconds.

Acid can cause severe eye and body damage. Operators should wear suitable face shields, rubber gloves, and boots. Usually eye fountains and showers plus adequate ventilation are provided. Extreme caution should be exercised when mixing acid with water—add slowly with constant stirring.

Following color treatments or cleaning and brightening, a protective coating may be desired to prevent staining, wear, or discoloration of the treated surface. Usually a colorless, transparent lacquer is used; these lacquers are compositions of thermoplastic resins dissolved in organic solvents with the resins generally belonging to the vinyl, acrylic, or cellulose family. Hardness and gloss depend on the formulation. The conditions of the intended application determine the requirements for the surface coating. Types of lacquers and the conditions for which they provide protection are listed below.

	Coating		
Application or Environment	Cellulose	Vinyl	Acrylic
Normal outdoor atmosphere	×	×	×
Marine environment		×	
Water immersion		×	
Chemical fumes		×	
Extreme sunlight		×	×
Abrasion		×	
Impact		×	
Marring		×	×
Flexing		×	
Acids		×	
Detergents			×
Staining			×
Gasoline	×	×	×

POLISHING AND BUFFING

The need for surface reflectivity or smoothness, deburred corners and edges, or freedom from dirt, oil, and other foreign substances may be satisfied in many instances by mechanical means. In this section we will first consider the "brute-force" processes, *polishing* and *buffing*, as means to achieve the required finish. This is followed by a review of inherent surface quality and the production methods used with different alloys which can influence these operations and may offer ways to reduce or eliminate them.

Sheet and strip surfaces of some metals can be quite rough because of the manner in which they are processed. Copper alloys exhibit these conditions to a lesser extent than some others, and therefore mechanical finishing of copper base materials can be done with comparative ease. The use of power brushing to mechanically remove surface irregularities is seldom needed, or is grinding, whereby 0.002 to 0.010 in. of metal is removed. *Brushing,* in copper alloy finishing, is used to impart a final brushed appearance and also to prepare the surface for other treatments like plating. Various kinds of wire or synthetic fiber are used for this purpose. Along with wheel speed and pressure, the type of brush used determines the final surface appearance.

POLISHING

Polishing is an abrading operation employed for the removal of grinding lines, scratches, pits, tool marks, and other minor surface defects. A belt or wheel containing a fine abrasive or used with an externally applied abrasive performs the polishing. Parts that are most readily adaptable to belt polishing are more than 0.5 in. in diameter and thicker than 0.025 in. and do not require more than 0.001 in. parallelism between the top and bottom surfaces. Most parts fabricated from copper alloy strip are polished using wheels. Polishing removes 0.0001 to 0.001 in. of metal to obtain a surface roughness of 16 in. or less. *Surface roughness (RMS)* is defined as

$$RMS = \left(\frac{h_1^2 + h_2^2 + h_3^2 + \ldots \; h_{n-2}^2 + h_{n-1}^2 + h_n^2}{n} \right)^{1/2}$$

where the peaks and valleys, H_n, are measured and recorded by a profilometer (see Figure 12–2); RMS is the root mean square of the height of the peaks and the depth of the valleys.

A surface roughness measurement can provide an approximate indication of the relative amount of polishing and buffing required,

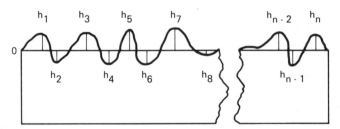

FIGURE 12–2
The profilometer traces and magnifies the surface profile and calculates the root mean square (RMS) of the peak deviations (h_n) from the base line.

although conditions other than surface roughness per se can also affect buffability.

Some operating variables that affect the efficiency of polishing and the quality of the worked surface are listed below.

1. Type and density of wheel. Polishing wheels can be made from several materials, such as canvas, muslin, leather, felt, or fiber. Disclike sections are sewn together and classified by hardness, as determined by the spacing of the stitching used to sew the pieces of material together. Wheels are classified by hardness and possible applications as follows:

Hardness	Spacing of Stitching, in.	Application
Very Hard	1/8	Flat surfaces, to level wavy surfaces
Hard	1/4	Same as above but with softer, less drawn out lines
Medium Soft	3/8	Curved, decorative parts that must not be flattened or leveled
Soft	1/2	For blending a curved surface and a feather edge
Very Soft	3/4	Limited use, for deeply curved surfaces, little surface work

2. Wheel preparation. For adhesive and abrasive material application, dressing the wheel precedes the addition of a hide glue or silicate base cement for attaching the abrasive material. This procedure insures a uniform working surface.
3. Wheel balance. Proper wheel balancing and good operator care and maintenance are required for good service life and effective operation.
4. Wheel speed. For brass and other copper alloys 4500 to 7500 surface ft/min is the recommended range.
5. Type of abrasive and grit size. For copper alloys either emery or aluminum oxide in a 180 or 220 grit size with oil is used. Endless cloth belts give polishing results similar to those of wheels but with the following advantage:
   a. Closer control of finish.
   b. Fewer polishing accessories.
   c. Less heat generation.
   d. Elimination of costly maintenance facilities.

BUFFING

Following a polishing operation, buffing is frequently used to produce a very smooth, mirrorlike, scratch-free surface. Cloth or sisal wheels suitably charged with a buffing compound perform this function by cutting or flowing the metal to eliminate minor defects and to impart a smooth, lustrous surface.

Buffing falls into two broad categories—hard buffing and color buffing. To subdivide these, the recommended operating variables shown below may be used. Hard buffing, which is sometimes called polishing, entails cutting the metal surface with a very fine grit cutting compound. The surface may have been polished by coarser grit compounds previously and require aggressive compounds and harder buffing wheels to obtain the desired smoothness. Color buffing produces the final lustrous finish, and softer, finer compounds and softer wheels are employed. Some buffing variables are:

Type of Buffing	Abrasive Compound	Wheel Hardness	Buffing Speed Surface, ft/min
Hard			
Satin finish	Aluminum oxide	Very Hard to Medium Soft	5000–9000
Cut-down	Tripoli	Hard to Medium Soft	5000–9000
Color			
Cut and color	Tripoli or coloring compound	Hard to Soft	5000–9000
Color	Vienna lime	Medium Soft to Very Soft	5000–9000

SPECIFYING SURFACES FOR POLISHING AND BUFFING

The brief discussion of finishing presented above gives the reader some general information about mechanical finishing processes. The purpose of this section is to enable the user to gain a better understanding of the surface characteristics of the copper and copper alloy sheet and strip furnished by brass mills for applications requiring special surface finishes such as buffing. As noted in Chapter 14, there are no published surface quality standards covering copper and copper alloy sheet and strip. Therefore, when purchasing these products, it is important that the buyer specify in the purchase order that the metal will be used in producing an article that requires buffing or any other special finishing requirement. This will allow the mill to select the best processing sequence to provide the required surface quality, as well as the mechanical and physical properties that are needed.

The control of surface quality needed for *buffability* starts with the casting process. Sound cast bars, free of porosity and inclusions, are required, and the continuous-type casting methods can provide such quality. With few exceptions, all copper and copper alloy sheet and strip processing includes a surface milling operation prior to the start of cold rolling. As described in Chapter 2, where exceptionally fine surfaces are required special care may be needed during the milling operation to insure that the milling cutters are sharp and the milled surface is free of any extraneous cutter marks. More rolling passes are frequently scheduled, along with additional annealing when necessary. At both intermediate and final rolling stages, rolling mills are thoroughly cleaned and inspected to be sure that surfaces will not be damaged. Oxidation must be minimized during annealing, and the metal must be thoroughly cleaned after annealing. Some or all of these special precautions may be required when the producing mill knows that the application requires metal with a buffable surface. These steps are taken as a means of eliminating minor surface blemishes that are not detrimental in applications where buffability is not required.

Even freedom from incidental defects is not necessarily synonymous with buffability. Other surface conditions also influence this characteristic. There is no standard test or measurement method to cover all aspects of buffability, so it cannot be completely evaluated, nor is it thoroughly understood. The mill exercises as much control as possible by selecting samples of metal which are buffed and examined during mill inspection.

Surface roughness is an element in the complex property of buffability and can be measured using a profilometer, as described earlier. While there are also other surface roughness measuring methods, RMS values are normally used when these requirements are specified. For metal that will be polished and buffed *with little or no forming* prior to finishing, surface roughness is more closely related to buffability. However, when a part is formed from the metal and subsequently polished and buffed, surfaces which are stretched 20% or more no longer reflect the original surface roughness. If the grain size is small enough, so no "orange peel" roughening develops, such surfaces will be smoother than the original metal. (Chapter 10 describes the effect of grain size on the surface roughness of drawn parts.) Conversely, the roughness of the original surface may be accentuated in areas of the formed part where compressive forces thickened the metal, as in the flange.

Parts with straight sidewalls can be ironed during drawing to produce smooth surfaces, but parts with curved contours will have some surface areas whose buffability reflects the original metal surface, and others that reflect the grain size of the metal.

A high degree of specular reflection has often been considered to be a measure of buffability, and there is no doubt that a mirrorlike metal surface without defects should be buffable. There are two methods of producing such a reflective surface. One is to use highly polished rolls and roll the metal without a lubricant. In such "dry rolling" only a very small reduction is possible, and a very slow speed is necessary to prevent overheating and damage. The metal must be paper interleaved to avoid friction scratches during coiling and uncoiling. In addition, only one temper can be produced, approximately 1/8 Hard. The cost of this process is generally prohibitive.

The second method is to use highly polished rolls and a water-soluble lubricant. Metal having a fine grain size is chosen, and a heavy rolling reduction is used to harden the metal, so its surface is less subject to damage. The metal surface will be burnished by the rolls and become quite reflective. If overheating is avoided, and the rolls do not become damaged, a smooth, bright metal surface can be produced. This metal can then be annealed to fine grain size and cleaned, so the gloss is not entirely lost. The finished metal is then suitable for shallow-drawn parts or flat parts that can be easily polished and buffed.

Highly reflective metal is more difficult to handle both by the parts fabricator and by the brass mill. The surface has a high coefficient of friction, galls more readily, and may become scratched in coiling and uncoiling. It has a low affinity for some lubricants, increasing the tendency for tool loading during drawing operations. Because of the difficulties in producing such metal in the brass mill, its high cost, and the tendency for extended delivery time, most mills prefer not to produce mirror-bright metal. Because of potential handling problems and the lubrication problems experienced by most parts fabricators, as well as the temper limitations, there is often no gain in total cost savings with such a product, even if buffing costs are minimized. The usual and more effective approach is to select a surface with a light, matte finish which can retain a protective film of lubricant. Then fewer parts are scrapped because of damage, and buffing costs are reasonably low. A defect-free, satin surface texture is the most desirable for a majority of brass applications in which the parts are finished by buffing. The automatic polishing and buffing machines used by volume producers, such as lockset manufacturers, are capable of producing beautifully buffed parts such as doorknobs and lock escutcheons. Doorknobs go through many press operations, and surface damage is avoided by the use of excellent tools and lubricants on metal surfaces which are not burnished but are smooth and defect free with a light matte texture.

The brasses—Alloy C26000, cartridge brass, in particular—are used for many drawn parts that are later finished. As noted in Chapter 10, for fabricating parts that are to be buffed, the product with the smallest average grain size that can be formed to the desired shape should be used to avoid surface roughening. When a small grain size is specified, the brass mill can use the ideal annealing practice to provide a smooth surface. Some alloys, such as C63800 and C68800, regularly have fine grain size and provide unusually smooth and bright surfaces on parts produced from them. It is sometimes possible to eliminate polishing and buffing entirely on parts which are finished by chrome plating when Alloy C63800 is used to replace C26000.

ROLLED PATTERNS ON STRIP

Embossed metal has some type of design or pattern impressed in its surface. Roll embossed strip usually fulfills a decorative or cosmetic requirement. In copper alloy sheet and strip this material is effectively produced by a final rolling process employing patterned rolls. This method provides a repetitive, longitudinal replica on one or both sides of the finished strip. Figure 12–3 shows one useful pattern of the many that are available from brass mills.

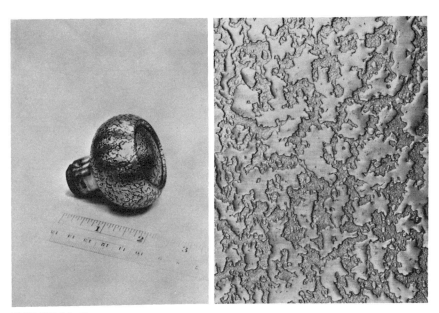

FIGURE 12–3
Photograph of surface of embossed metal. The darker areas represent the recessed area, and the light areas the raised surface. Magnification: approximately 2x.

The embossed antique pattern of Figure 12–3 typically is impressed to a depth of 0.001 to 0.005 in. and is available in an "as rolled" or reannealed condition. Most material for this product has a grain size around 0.040 mm before embossing. The "as rolled" strip undergoes slight work hardening equivalent to 1/8 to 1/4 Hard temper and is suitable for many shallow-drawn and formed parts. More severe forming usually requires annealing the embossed metal to remove residual stresses and to obtain added ductility. It should be noted that the original grain structure does not fully recover and recrystallize but exhibits some grain growth.

The knurled material of Figure 12–4 may be manufactured in rolled or annealed tempers, but the typical strip product, cold rolled with a diamond pattern depth of approximately 0.010 in., is normally used in the "as rolled" condition.

Besides the use of the antique patterned strip to produce doorknobs, plumbing hardware, and other household fixtures, such strip is also readily formed into welded tubing. Embossed tube is also used in decorative applications. Embossed with a specific pattern, (e.g., ribs or fluting), the tube has better heat transfer capabilities than smooth surface tubes and is used for some heat exchanger applications. The

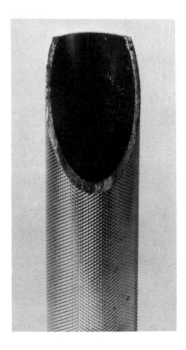

FIGURE 12–4
Photograph of knurled finish. The diamond pattern is embossed to a depth of about 0.010 in.

patterned surface also provides better gripping action, such as might be required when plastic inserts are assembled with the tubing.

Other applications in which copper alloys are embossed include coined products, typified by monetary coins, medallions, and key blanks. The metal usually receives an initial blanking operation for these parts and is then coined under tremendous pressure to impart the desired grooves, lettering, and other designs. Generally, soft, large-grained metal must be used to allow this extreme forming and metal flow. Rolled temper may be used for blanking, and the blanks annealed and cleaned before embossing. More data on coining can be found in the section entitled "Miscellaneous Forming Operations" in Chapter 10.

BIBLIOGRAPHY

1. Kihlgren, T. E., N. B. Pilling, and E. M. Wise, *Trans. AIME,* **117,** 279–309 (1935).
2. *Metals Handbook,* Vol. 2, American Society for Metals, Cleveland, 1964.

13

SPRING DESIGN AND MATERIAL SELECTION

Because of a favorable combination of strength, conductivity, formability, and corrosion resistance, copper alloys are often selected for springs. Copper alloy springs are designed following the same general engineering rules applied to any other metal family. However, there are a few areas where coppers are different and special attention is warranted. The purpose of this chapter is to review both the classical engineering methods and the special aspects that apply to copper alloys.

The chapter will cover, in turn, design criteria, properties of copper alloy strip that influence the selection of spring material, and application of property data to designing springs typified by the flat strip which is fixed at one end and has a deflecting force applied at the other end. The discussion of properties includes three main factors: mechanical requirements, formability, and cost. The section on applications includes testing, a selection model, and example problems.

DESIGN CRITERIA

Any spring is designed to perform a particular and specific function. Satisfactory achievement of that function depends upon the spring's having the correct geometric form and mechanical properties. The mechanical properties of the spring are directly related to the properties of the metal from which it is fabricated.

Only with a thorough understanding of a spring's function can the designer determine the best geometry and proper material. Unrealistic definition of what a spring must accomplish can result in over design and an excessively expensive part, or an underdesigned part that may fail in service. Experience has shown that careful initial statement of these requirements can save a great deal of time, money, and frustration.

Design parameters fall into three broad categories:

1. Mechanical (physical, corrosion and mechanical).
2. Formability.
3. Cost.

While the first and the last are obvious, the formability criterion may seem out of place as a part of design considerations. But no matter how careful the choice of material from the standpoint of mechanics and cost, the effort is wasted unless the part can be satisfactorily fabricated.

Included in mechanical criteria to be considered are such factors as amount of deflection, contact force or pressure, contact resistance, minimum number of life cycles, and nature of the environment. Formability includes bending and coining. Cost criteria include metal cost, tooling, plating, and heat treating.

Not only must all these parameters be defined, but they must also be assigned priorities. The final selection of a material and geometry often requires some compromise between the criteria in the extent to which each can be satisfied.

PROPERTIES

This section will be concerned with design requirements and related material properties. As springs are basically mechanical devices, the mechanical requirements will be discussed first. It should be kept in mind that various properties are interrelated and cannot be considered independently.

Mechanical Criteria
DEFLECTION. The purpose of any spring is to store energy transmitted to it by deflecting or straining and then to release the energy when allowed to return to its initial position. The energy is stored as strain-energy (in.-lb/in.^{3}) within the material. If the strain becomes too great, the stress will exceed the elastic limit, and plastic deformation or permanent set will occur. The part then will not return to the original free position, and the spring has "failed." As noted in Chapter 5, the elastic limit is not a readily measured property, and some fraction of the 0.2% offset yield strength generally replaces it in design calculations.

It is essential to know how much deflection is required. With this knowledge, the stress level in the part can be calculated using the techniques described later in this chapter. The stress can then be compared to the yield strength of the material to determine if the material is strong enough not to be permanently deformed.

Published yield strength data for cold-rolled copper alloy strip have generally included only tensile yield strength measured on specimens taken with the axis parallel to the direction of rolling. When the same materials are tested in compression, the 0.2% offset yield strength is found to be considerably lower. Table 13–1 gives tensile and compressive yield strengths of two alloys in two different tempers and illustrates such differences. The compressive yield strengths in these cold-rolled materials are consistently lower than the tensile yield strengths in the longitudinal direction. The fact that the residual stresses from rolling are predominately compressive in the longitudinal direction probably accounts for much of this difference.

TABLE 13-1
Effect of Stress Mode on 0.2% Offset Yield Strength

Alloy	Temper	Tensile Mode		Compressive Mode	
		ksi	Mpa	ksi	Mpa
C51000	Hard	81	558	62	427
C51000	Spring	111	765	82	565
C68800	½ Hard	72	496	48	330
C68800	Spring	112	772	61	421

Property differences between specimens tested in tension and compression are also evident when specimens are taken so the axis is transverse to the direction of rolling. Table 13–2 gives yield strengths in the transverse direction for three of the materials listed in Table 13–1, as

TABLE 13-2
Effect of Stress Mode and Orientation on 0.2% Offset Yield Strength

Alloy	Temper	Orientation	Tensile Mode		Compressive Mode	
			ksi	Mpa	ksi	Mpa
C51000	Hard	Longitudinal	81	558	62	427
		Transverse	77	531	87	600
C51000	Spring	Longitudinal	111	765	82	565
		Transverse	107	738	122	841
C68800	Spring	Longitudinal	112	772	61	421
		Transverse	103	710	120	827

well as repeating the longitudinal values for comparison. In the transverse direction the yield strengths in compression are higher than those determined in tension. These kinds of differences are regularly observed in cold-rolled copper alloy strip.

The stress applied to flat springs or formed springs made from copper alloy strip is generally such that bending or flexure occurs. There is a compressive stress component as well as a tensile stress component. Figure 13–1 illustrates this schematically. In symetrical sections, these two components can be assumed to be approximately equal. Therefore, when the spring designer uses the yield strength as a measure of the stress at which permanent deformation will occur, the average of the tensile and compressive yield strengths is a more appropriate value than the tensile yield strength frequently used.

Further more, the designer may well take the orientation into consideration, because the average of yield strengths in the transverse direction is higher than that in the longitudinal direction. Therefore springs designed so the flexure is in the transverse direction can withstand greater stress before permanent deformation occurs than springs stressed in the longitudinal direction.

CONTACT FORCE.   Whenever a spring is deflected, it offers an equal and opposing or reactive force to the applied stress. This *contact force* increases proportionately with deflection until the yield strength is approached. The magnitude and predictability of this force are of utmost concern to the designer of springs used in electrical contacts.

Contact force is a function of several factors, including metal thickness, length, and stiffness. The classical measure of a material's stiffness when stressed below the proportional limit is Young's modulus. In recent years it has been recognized that a *secant modulus* based on the

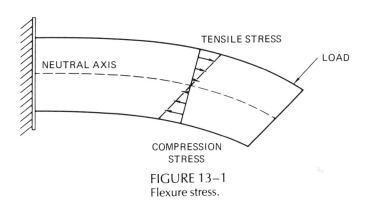

FIGURE 13–1
Flexure stress.

predicted stress is more useful and realistic and can be applied to copper alloys stressed above the proportional limit. More information on the secant modulus is given in Chapter 5.

When the concept of secant modulus is used, it is necessary to recognize that the stiffness of a material varies not only with alloy, temper, mode, and orientation, but also with stress level. Stiffness, as a function of these factors, is presented in a series of curves such as those in Figure 13–2. The horizontal portion of the curves are Young's moduli for the materials.

In order to select the correct modulus value the designer must know the alloy, temper, mode, and orientation. This allows selection of the appropriate applied stress-modulus curve.

Having such a curve, there are then two techniques for determining the secant modulus, which will permit more accurate calculation of the stress associated with a certain deflection. In the first method, beginning with a given deflection and using Young's modulus as an approximation, the stress is calculated. This value is then used to obtain a secant modulus value from data such as those plotted in Figure 13–2. This, in turn, is substituted in the equation, and a second stress calculated. A second

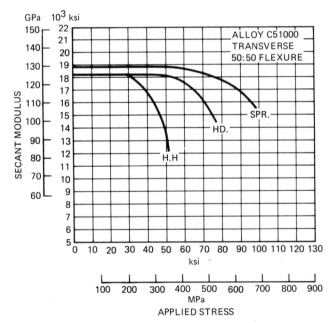

FIGURE 13–2
Variation of the secant modulus of alloy C51000 with applied stress and with various tempers.

secant modulus is read off the chart, and a third stress is then calculated. With each reiteration the calculated stress becomes more accurate until, after several repetitions, final stress and modulus values are established, which are very close to the actual values.

The second technique is graphical. First, a modulus is calculated for two arbitrary stresses, using the classical linear equation for a cantilever beam which relates stress, beam dimensions, and modulus. These calculated values are then plotted on a graph of measured secant modulus values as shown in Figure 13–2 and connected with a straight line. The intersection point between this line and the measured line provides the corect secant modulus and stress for the particular deflection. The values can be used for contact force calculations and for indicating the yield strength that will be needed in the material for the particular spring geometry. An example showing the application of these techniques is given at the end of this chapter.

Over a period of time, contact force decreases for a spring held at constant deflection. This is the result of *stress relaxation*, a property of metals which is discussed in Chapter 7. Stress relaxation data are particularly important to spring design since they are needed to predict the contact force over the expected life of the part. Figure 13–3 shows a plot of stress relaxation data for Alloy C72500 in Spring temper at room temperature (RT) and at 105°C (221°F).

Observe the effect that temperature has on the stress relaxation rates. This effect is an important consideration for springs operating at temperatures above room temperature, as discussed later in the section on the effects of environment.

Stress relaxation data can be used in two ways. The designer can establish the criteria (alloy, temper, mode, etc.); calculate the stress (using the secant modulus); then, using the appropriate stress relaxation curve, determine whether or not the stress remaining at the end of the required time will produce adequate contact force.

Instead, knowing the contact force requirement over a given time period, the designer can "back into" the correct material. This is done by searching the relaxation curves for one that appears to meet the long-term requirements. Contact force calculations using the corresponding stress level are made, and if the result is adequate, the final mode, orientation, alloy, allowable stress level, and temper are established. It is then necessary to calculate what deflection will generate the necessary initial stress and contact force.

The first technique is used when certain factors, such as mode and orientation, are pre-established. However, when the design has no pre-established limits, the second technique is more direct. A sample problem, using both approaches, can be found at the end of this chapter.

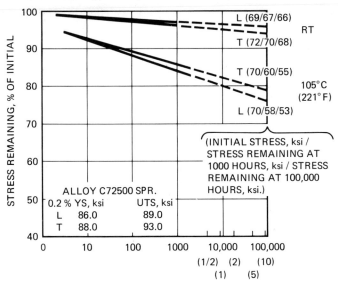

FIGURE 13-3
Stress relaxation data in flexure mode at room temperature (RT) and 105°C (221°F) for Alloy C72500, Spring temper, longitudinal (L) and transverse (T) directions.

Often contact force is important as the spring must "grip" another component, as in many electrical connector applications. One method of judging *grip strength* is the insertion-withdrawal test. The results from this type of test are also affected by a surface characteristic, the *coefficient of dry, sliding friction*. The mathematical relation governing the grip strength of two surfaces is

$$F = \mu N$$

where F = grip force (force to withdraw)
μ = coefficient of dry sliding friction
N = normal contact force

When more than two surfaces are involved, $F = \Sigma \, \mu N$ for each pair.

Table 13-3 gives typical values of μ for some copper alloys. Surface coatings, such as tin, solder, or gold, will change the coefficient.

Contact resistance is a characteristic often associated with contact springs which function as current-carrying members. As discussed in Chapter 10, contact resistance is highly dependent upon contact force. The effects of stress relaxation are important here, for as the contact force decreases, the contact resistance will probably increase.

TABLE 13-3
Coefficients of Dry Sliding Friction

Alloy	Soft	Hard
C66400	0.224 ± 0.023	0.167 ± 0.024
C41100	—	0.184 ± 0.022 (Spring)
C26000	0.397 ± 0.045	—

LIFE CYCLES. Springs must be capable of withstanding repeated loadings without failure. The measurement for this characteristic is *fatigue strength*. An example of a potential fatigue problem is a relay spring that is repeatedly deflected to make electrical contact, then returned to a free position. While the stress generated by each deflection is below the yield strength, the part may eventually crack and fail from fatigue. The designer, to predict how many cycles the material is capable of withstanding, must know the maximum stress that may be applied over any given number of loadings. For a detailed discussion of fatigue refer to Chapter 7.

Fatigue data for design purposes are presented in the form of a stress-number of cycles, or *S-N curve*. Such a curve is shown in Figure 13–4. This type of curve is particularly useful, but takes considerable testing time to develop.

Tabular data, as shown in Table 13–4, are useful for quick, qualitative comparisons between different alloys and tempers. A standard number

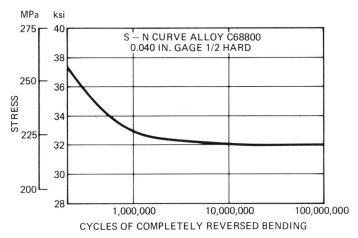

FIGURE 13–4
S-N curve for Alloy C68800 in ½ Hard temper.

TABLE 13-4
Fatigue Strength Data

Alloy	Temper	0.2% Yield Strength		Fatigue Strength— 10^8 Cycles	
		ksi	Mpa	ksi	Mpa
C19400	Spring	67.9	468	21.8	150
C26000	Spring	83.8	578	26.6	183
C42500	Extra Spring	86.2	594	36.0	248
C51000	Spring	103.3	712	34.1	235
C63800	Hard	106.0	731	41.0	283
C68800	Spring	101.5	700	34.5	238
C76200	Spring	108.3	747	30.0	207

of cycles is chosen, usually 100 million, and the maximum allowable stress for this number of cycles is determined.

Fatigue strength, like yield strength, for cold-rolled tempers is dependent upon orientation with respect to rolling direction. Data, particularly S-N curves, are not readily available for the various orientations as yet. However, test results on longitudinal specimens are available, and 45° and transverse data are being generated. Table 13–5 gives some data on the effect of orientation with respect to rolling direction on fatigue.

ENVIRONMENT.   Few springs ever operate in a perfect environment. In fact, the environmental aspects are often the most difficult to deal with.

Corrosion.   Of particular concern is corrosion. Corrosion of copper and its alloys is discussed in detail in Chapter 8. Basically, copper alloys are quite resistant to attack by most types of atmospheres, with the exception of a few that are susceptible to stress corrosion, or season cracking. This

TABLE 13-5
Effect of Orientation on Fatigue Strength

Alloy	Temper	Orientation to Rolling Direction	0.2% Yield Strength		Fatigue Strength— 10^8 Cycles	
			ksi	Mpa	ksi	Mpa
C26000	Extra Spring	Longitudinal	82.9	572	27.0	186
		45°	86.1	594	29.5	203
		Transverse	92.9	641	37.5	259

type of corrosion occurs only in atmospheres containing unusually aggressive compounds but can have catastrophic effects on spring performance.

Stress corrosion occurs when susceptible materials, containing unequally distributed stresses, are in one of these aggressive environments. By their very nature springs are subject to nonuniform stresses. Typical hostile environments include ammonia vapor, salt spray, and mercury compounds.

Copper alloys containing zinc in an amount greater than 15% are the most susceptible to stress corrosion. The relative resistance ratings based on laboratory tests of several copper alloys to stress corrosion are shown in Table 13–6. Ratings are on a scale based on a low resistance rating of 10 for 70/30 brass, Alloy C26000, and of 1000 for very resistant alloys.

Temperature. Temperature effects on copper alloy springs are important for two reasons: (1) often these springs must function at elevated or depressed temperatures; (2) elevated temperatures are sometimes used as a way of shortening time in tests to evaluate the life of a copper alloy product.

Since copper and its alloys have face-centered cubic lattices, they are not subject to the *ductile-brittle transition* at low temperatures that is common to body-centered cubic metals such as steel. However, copper alloys may still be subject to stress relaxation even at low temperatures. Relaxation rates will show a moderate decrease at low temperature because of the lower thermal energy and the subsequent slowing of dislocation movement. It is at elevated temperatures that copper alloy spring performance is most affected.

TABLE 13-6
Relative Stress Corrosion Resistance Ratings
of Several Copper Alloy Spring Materials

Alloy	Resistance Rating
C26000	10
C19400	Immune
C19500	Immune
C51000	1000
C63800	500
C68800	100
C72500	1000
C76200	30
C77000	50

Thermal softening may occur because the temperature of exposure is sufficient to cause partial or complete annealing of the metal. Annealing is discussed in depth in Chapter 3. Table 13–7 shows the effect of temperature on the tensile strength and yield strength of various spring alloys after exposure up to and including the temperature at which at least partial recrystallization has occurred.

The data in Table 13–7 are useful for making qualitative comparisons of various alloys. Note that the strength of some alloys is increased by exposure to subrecrystallization temperatures. When the temperature is above the recrystallization point, strength decreases rapidly.

Softening is a problem where heat is applied during fabrication or assembly, such as a spring subjected to welding or brazing. Also, operations such as sintering, encapsulating, and molding assemblies can soften the base metal. If excessive softening occurs, the strength of the spring will be decreased to the extent that it will not function effectively.

A more common problem is the marked increase in relaxation rates of loaded springs when the operating temperature increases. Figures 13–3 and 13–5 show the effect of temperature on the relaxation rates of two different alloys. The magnitude of the change in relaxation rate with temperature increases is not constant but depends upon many factors (alloy, temper, stress level, temperature, etc.); however, the rate of stress

TABLE 13-7
Softening Resistances of Spring Temper Copper Alloys

Alloy	Before Exposure[a]	Temperature,[b] °C (°F)				
		100 (212)	200 (392)	300 (572)	400 (752)	500 (932)
C19400	74/71	74/71	72/68	70/63	48/21	46/20
C19500	88/85	88/85	86/84	83/77	73/68	62/39
C26000	92/81	92/83	96/90	65/41	57/28	50/17
C42500	93/88	93/88	92/88	84/79	54/35	51/28
C51000	102/100	102/100	98/94	89/80	57/34	54/25
C63800	132/116	134/119	135/122	132/120	86/63	84/55
C68800	126/115	127/118	139/135	115/105	84/57	82/55
C72500	100/98	100/98	101/99	98/97	103/98	91/81
C76200	114/112	114/112	120/119	1w1/119	84/71	74/48

[a]Tensile strength and 0.2% yield strength, ksi. Mpa values can be obtained by multiplying by 6.895.
[b]Exposure is for 1 hour at temperature indicated. The tensile tests were performed at room temperature after cooling from the elevated temperature given.

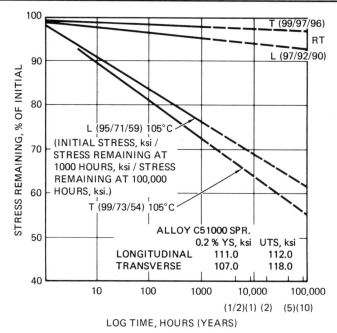

FIGURE 13-5
Effect of elevated temperature on stress relaxation rate of Alloy C51000.

relaxation does increase for all copper alloys as the temperature increases. This lack of a constant relationship between temperature, time, and relaxation rate is particularly important if the designer is using elevated temperature to accelerate a life test. This topic is discussed further in the section entitled "Testing".

If a spring must function at elevated temperature, and provide a certain level of contact force, the spring designer needs pertinent stress relaxation data. Those data can be used in a manner similar to that described earlier under contact force. If the data are not available for the exact set of circumstances, the material supplier should be able to provide qualitative judgments that will permit a judicious selection of a suitable material based on approximations or extrapolations of data that are available.

Formability

The most careful and thorough material selection process for functional properties is worthless if the part fractures during fabrication. Spring materials are usually subjected to one of two basic fabrication operations: coining and bending.

COINING. *Coining* is the term applied to impact-cold reduction of metal. An in-depth discussion of this topic can be found in Chapter 10. This process increases the "strength" of the metal as well as decreasing the thickness. The effect that coining will have on strength (and subsequent formability) can be predicted using a percent reduction by cold-rolling curve for the alloy in question. Figure 13-6 is a typical curve, in this case for Alloy C68800.

An example of the effect that coining has on strength is a part made from ½ Hard Alloy C68800, which is coined from 0.025 in. thick to 0.013 in., a 48% reduction. The metal to be applied is certified with a tensile strength of 102 ksi. From Figure 13–6, 21% reduction by cold rolling produces 102 ksi.

Then it follows that the RF (annealed and ready to finish) gauge (the gauge before the final 21% reduction) was

$$\text{RF Gauge} = \frac{0.025}{(1-0.21)} = 0.0316 \text{ in.}$$

Then, from the original annealed condition, the total work in the coined region is

$$\text{Percent Total Cold Work} = 1 - \left(\frac{0.013}{0.0316}\right) \times 100 = 58.9\%$$

Again, using Figure 13–6, the final properties in the coined area will be

Tensile strength	130 ksi	895 Mpa
0.2% Yield strength	117 ksi	805 Mpa
Elongation	2%	

This estimate of the strength of the reduced section can now be used for bend analysis, stress relaxation, deflection, and contact force calculations. The same considerations regarding mode and orientation of the stress application apply here. One point to be noted is that coining induces unequal stresses in a member. This can make it more susceptible to stress-corrosion cracking.

BENDING. Generally, springs fabricated from strip are formed into configuration by *bending.* Some of the information on bending given in Chapter 10 is repeated here for convenience of the reader. Figure 13–7 shows the geometry of a typical 90° bend. In addition to thickness, bend radius, *direction of grain* (rolling direction), and bend angle, the alloy and

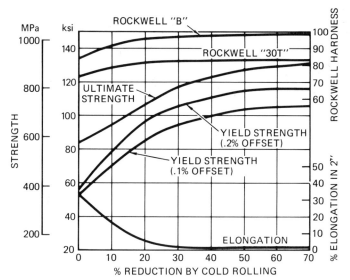

FIGURE 13–6
Effect of varying amounts of cold work on the hardness, strength, and ductility of Alloy C68800.

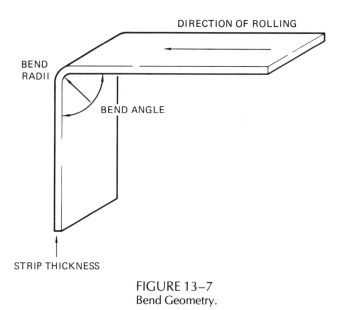

FIGURE 13–7
Bend Geometry.

temper are important. Of particular significance is the orientation of the bend axis with respect to the direction of rolling of grain. Figure 13–8 shows "good-way" and "bad-way" *bend orientation*. Bends made at angles other than 0° and 90° to the rolling direction should be considered to have intermediate formability—the radius over which the bends can be made without fractures occurring is usually greater than that required for good-way bends but less than that required for bad-way bends.

As a general rule, bad-way bends require more generous bend radii than good-way bends on the same spring material. However, from earlier discussions it will be recalled that optimum strength is obtained when loading is in a transverse orientation. The designer sometimes must work with bad-way bends in order to optimize performance.

Thickness, temper, bend directions, and bend radii for bending an alloy are related by a family of *bend curves*. Figures 10–23 and 10–24 are typical curves showing for various thicknesses the recommended *minimum bend radii* for Alloy C51000 strip.

In order to use these curves the designer must first establish certain variables. The alloy, thickness, and approximate temper are generally chosen based on functional requirements of the part. Of the remaining criteria, direction of bend and radius of bend may be varied to some extent to satisfy formability and cost requirements.

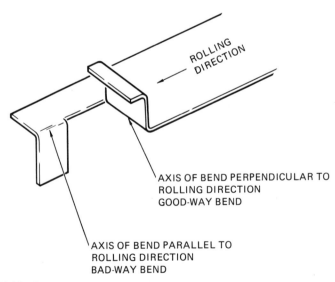

ROLLING DIRECTION

AXIS OF BEND PERPENDICULAR TO ROLLING DIRECTION GOOD-WAY BEND

AXIS OF BEND PARALLEL TO ROLLING DIRECTION BAD-WAY BEND

FIGURE 13–8
Orientation relative to rolling direction determines good-way and bad-way bending.

Following are brief problems to illustrate the use of these bend data.

A. SITUATION: Known Alloy C51000, bend is good-way, metal thickness must be 0.015 in., temper must be Extra Hard in order to meet mechanical requirements. Determine the minimum radius to form part.

   SOLUTION: Select Alloy C51000, good-way curves as shown in Figure 10–23. Follow the Extra Hard curve to the intersection with the horizontal line corresponding to 0.015 in. thickness. The vertical line passing through this point is the *minimum* bend radius that can be used—in this case, 0.005 in.

B. SITUATION: Known Alloy C51000, bend is bad-way, metal thickness must be 0.030 in. Further, tools are already complete, radius established as 0.050 in. Determine the maximum temper that will successfully fabricate the part.

   SOLUTION: Select Alloy C51000, bad-way curves as shown in Figure 10–24. Find the intersection of the horizontal line passing through a thickness of 0.030 in. and the vertical line drawn through corresponding to a radius of 0.050 in. The nearest temper to the left and above this point represents the maximum temper that is usable—in this case, Hard.

Because of the overlap of tempers and properties it is important to consider tensile strengths as well as temper names when dealing with bend curve data. It is possible that the low-tensile-strength end of Extra Hard temper could be used to fabricate the part in Example B Also, if a particular bend is not a critical or functional bend, a slight fracturing may be tolerated in order to satisfy a spatial or geometric requirement.

Cost

Chapter 15 is concerned with a variety of factors important to cost. This section will consider cost as it applies specifically to spring design and material selection.

The first factor, of course, is the cost of the base metal. Table 13–8 shows the relative costs for alloys most often used for springs. The values have been adjusted to allow for differences in density, so that the index reflects only the cost of strip to provide a given number of pieces.

Cost will be affected by factors such as thickness tolerances closer than standard. Similarly, the thinner and narrower the strip, the greater is the cost. The choice of strip width can be affected by the choice of fabricating equipment. In multislide presses, for instance, the part is usually laid out parallel to the rolling direction. As a result, the strip is used in relatively narrow widths. If the same part is to be stamped on a progressive die, it is usually placed perpendicular to the direction of rolling and uses wider strip. While the tools for multislide presses are generally less expensive, the lower cost of wider strip can make up the difference over a period of time. Another advantage is that parts cut out transverse to the rolling direction often give superior mechanical performance. Chapter 15 contains additional information on the influence of blank layout on cost.

Blanking and piercing tools should be resharpened after running a certain number of strokes. Certain materials are more abrasive to tools than others. Phosphor bronzes, unless carefully processed, can be particularly abrasive, and beryllium-bearing alloys can also be abrasive unless properly prepared for pressworking. This is also true for aluminum-bearing alloys, if not cleaned to remove aluminum oxides. If the spring is small in size or intricately shaped, and run in high volume, tool wear cost can be particularly significant.

Coatings can also have a dramatic cost impact. If it is desired to improve the base metal corrosion resistance or contact resistance performance, or if the metal has an abrasive surface, it may be necessary to apply a coating. In some cases the added cost of this coating can be eliminated by switching to a different base material.

TABLE 13-8
Relative Cost Factors for Various Spring Materials

Alloy	Cost Factor[a]
155	1.23
194	1.15
195	1.26
175	3.14
260	1.00
425	1.19
510	1.52
638	1.09
688	1.03
725	1.50
752	1.58
770	1.47

[a]Based on 5000 lb of .025 × 1.000 in. strip, adjusted for density. Alloy C26000 set as base as of August 1976.

APPLICATION

The preceding sections dealt with basic design and performance criteria. This section describes the procedures for applying these criteria in order to optimize material selection.

Selection Model

In order to keep all of the various design requirements in perspective, and thus to always select the optimum material, the designer should have a consistent and logical material selection model. The following is one such model which has been found effective. This system has proved a useful tool over several years of application.

Using this procedure with an already chosen design will certainly result in adequate material selection. However, the optimum results can be obtained if the design and the material selection are conducted with some interplay between the two in the designer's mind while the initial concept is being established.

1. Application of the model can start once the basic geometric and dimensional parameters have been established.
2. The designer then lists all of the necessary mechanical criteria, including such factors as conductivity, contact force, fatigue life, and corrosion environment. Priorities are assigned to the criteria.
3. Forming requirments are established. These encompass such factors as coining requirements, bend radii, and bend orientation.
4. The designer selects several alloys for comparison. This initial selection can be aided, of course, by consulting potential suppliers.
5. On the basis of the coining and alloy bend data, an optimum temper for each alloy is selected.
6. These various alloy-temper combinations are next analyzed to see if they meet the mechanical requirements.
7. If one or more alloy-temper combinations meet these requirements, the designer can continue in the selection process. If none of the first selections is satisfactory, the designer must choose another group of alloys for trial or modify some of the pre-established criteria. Then the process is begun again.
8. Once the formability and mechanical criteria are met, the designer can compare the selected alloys and tempers against the remaining criteria, such as cost and availability.
9. Sample material can now be procured, sample parts fabricated, and testing begun. If the parts do not perform satisfactorily (assuming the test procedure is valid—see the section in this chapter entitled "testing"), the designer must begin the process again, this time carefully reevaluating the original design parameters.

Throughout this process the designer will find it to his advantage to work in close conjunction with the technical sales staffs of potential material suppliers. A flow chart for this model is shown in Figure 13–9.

Computer Assist

The designer can be greatly aided by the use of a computer in the design and material selection process. Parts with complex geometries can require sophisticated mathematical analysis that can be time consuming if done manually, especially if the calculations must be repeated.

Programs are available commercially from numerous sources, either on a purchase or rental (time share) basis. The programs are usually based on finite element analysis and are capable of handling virtually any configuration. Typical programs will give stress level data, load-deflection data, and contact force data.

Most commercial programs are arranged in such a manner that they are easy to learn and use, and most do not require a knowledge of computer programming. A spring design program should be selected that allows the use of secant modulus data.

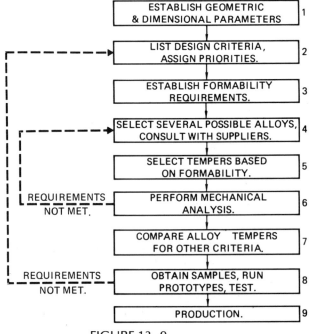

FIGURE 13–9
A material selection model.

Sample Problem

The following sample problem shows the application of the secant modulus for design purposes. The problem is used by courtesy of Mr. S. Paul Zarlingo and is taken, in part, from his paper entitled "Predicting Contact Force, Deflection, and Stress in a Terminal/Connector Design" (see Reference 6).

The classical beam formulas are based upon geometry ("beam" length, width, thickness) and the inherent stiffness of the material (Young's modulus of elasticity). Numerous applications reduce to the simple cantilever beam, fixed at one end and loaded at the opposite end (Figure 13–10).

Contact Force and Deflection

Contact force is of paramount interest because surface films, usually of high resistivity, must be "broken." From Figure 13–10, equation (2),

$$P = \frac{WT^3ED}{4L^3}$$

Assume that a contact force of 250 g (0.551 lb) is desired and that Spring temper phosphor bronze Alloy C51000, a popular high quality copper

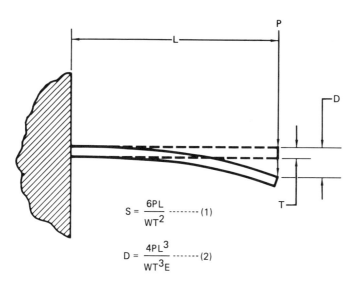

$$S = \frac{6PL}{WT^2} \ \text{------(1)}$$

$$D = \frac{4PL^3}{WT^3E} \ \text{------(2)}$$

S = MAX. STRESS; P = LOAD; E = MODULUS OF ELASTICITY; W = WIDTH; L, D, T AS SHOWN

FIGURE 13–10
The cantilever beam.

alloy spring material, is to be used. Let the geometry be determined; for example, $W = 0.100$, $T = 0.020$, in. $L = 1.00$ in. Young's modulus for Spring temper C51000 is 16.8×10^6 psi. Thus, classically, for deflection, Dym,

$$0.551 = \frac{0.100 \times 0.020^3 \times 16.8 \times 10^6 \times Dym}{4 \times 1.00^3}$$

$$Dym = 0.164 \text{ in.}$$

But, if the stress level is above the proportional limit, Young's modulus does not apply and the deflection of 0.164 in. will be an inaccurate prediction (see Chapter 5).

Maintaining the desired contact force, we can predict the maximum stress. From Figure 13–11, equation (1)

$$S = \frac{6 \times 0.551 \times 1.00}{0.100 \times 0.020^2}$$

$$S = 82.5 \text{ ksi}$$

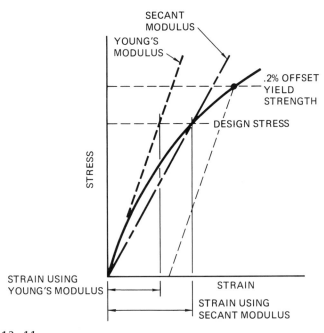

FIGURE 13–11
Typical stress-strain diagram showing the difference between deflections determined from Young's modulus and from secant modulus.

The tensile strength of Alloy C51000 for Spring temper is 95 to 110 ksi; the corresponding 0.2% offset yield strength is 92 to 108 ksi. The maximum operating stress just calculated is 82.5% of the nominal yield strength! At this high stress level the *secant modulus*, which is a function of stress level (as shown in Figure 13–12), must be used to predict the deflection. Figure 13–12 shows quantitatively the relationship between secant modulus and stress. Note that at the calculated stress level of 82.5 ksi the secant modulus is 13.5×10^6 psi. Young's modulus, previously used to calculate deflection, is 16.8×10^6! Thus, the deflection is more accurately determined from

$$0.551 = \frac{0.100 \times 0.020^3 \times 13.5 \times 10^6 \times D_{sm}}{4 \times 1.00^3}$$

$$D_{sm} = 0.204 \text{ in.}$$

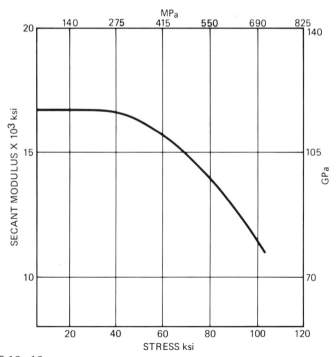

FIGURE 13–12
Variation of secant modulus ($E\hat{s}m$) with stress. Alloy C51000, Spring temper, bending mode, longitudinal direction.

The previous calculation, $D_{ym} = 0.164$, in., has an error of 20%!

An even greater error is introduced if Young's modulus is used for calculating deflections at higher stresses. From Figure 13-10, substituting equatrion (2) in equation (1) gives

$$S = \frac{3TED}{2L^2}$$

$$S = \frac{3 \times 0.020 \times E \times D}{2 \times 1.00^2}$$

$$S = 0.030ED$$

Classically, then, at a stress of 100,000 psi (the nominal yield strength of Spring Temper Alloy C51000),

$$100{,}000 = 0.030 \times 16.8 \times 10^6 \times D_{ym}$$

$$D_{ym} = 0.198 \text{ in.}$$

But, using the secant modulus of 11.5×10^6 psi (from Figure 13–12),

$$D_{sm} = 0.289 \text{ in.}$$

In this example, the first deflection calculated contains 31% error!

Genralized Load-Deflection Relationships

For a given geometry, stress can be expressed as

$$S = kED; \; k = \frac{3T}{2L^2}$$

In the given cantilever

$$k = \frac{3 \times 0.020}{2 \times 1.00^2}$$

$$k = 0.030$$

Thus, as shown before,

$$S = 0.030ED$$

$$D = \frac{100S}{3E}$$

From Figure 13–10, stress can also be expressed as a function of load (contact force):

$$S = \frac{6PL}{WT^2}$$

$$S = \frac{6 \times 1.0 \times P}{0.1 \times 0.02^2}$$

$$S = 150,000P; \quad P = \frac{S}{150,000}$$

With the above equations and Figure 13–12, the relationship between load and deflection can be established.

S, ksi	D_{sm}	E_{sm} (graph)	P lb	g
30	0.060	16.8	0.200	91
40	0.081	16.5	0.267	121
50	0.103	16.2	0.333	151
60	0.127	15.7	0.400	181
70	0.156	15.0	0.467	212
80	0.193	13.8	0.533	240
90	0.240	12.5	0.600	272
100	0.290	11.5	0.666	302

Note that these data are compiled utilizing modulus as function of stress (as it is at high stresses). Classically, in determining load-deflection relationships, modulus is considered constant. Thus, from Figure 13–10, equation (2),

$$D = \frac{4 \times 1.00 \times P}{0.100 \times 0.020 \times 16.8 \times 10^6}$$

$$D = 0.288P$$

The tabulated data (using the secant modulus) and the straight line equation above (using Young's modulus) are graphically depicted in Figure 13–14. At low stress levels, the classical use of Young's modulus does not introduce error (secant modulus equals Young's modulus). However, at high stress levels, the secant modulus must be used in order to avoid significant error in predicting load or deflection.

Stress and Deflection

In the previous analysis, attention was focused on contact force and the deflection was viewed as a subsequent event. In numerous applications, however, the deflection is "predetermined" by constrictions on the connector, for example, location in a plastic slot. The connector's deflection is arrested by the plastic stop. In such cases, the stress developed due to the deflection, is the desired calculation. As before,

$$S = \frac{3TED}{2L}$$

With a "predetermined" deflection, for example, 0.200 in. (geometry as before),

$$S = \frac{3 \times 0.020 \times 0.200 \times E}{2 \times 1.00^2}$$

$$S = 0.006 \times E$$

Using Young's modulus, $E_y = 16.8 \times 10^6$ psi, gives

$$S = 0.006 \times 16.8 \times 10^6$$

$$S = 101 \text{ ksi}$$

But classical methods may yield erroneous values for stress just as they did for deflection. In this case, the "classical" stress is at the yield strength of the material. For such conditions, Young's modulus cannot be used to calculate the stress; gross error will result.

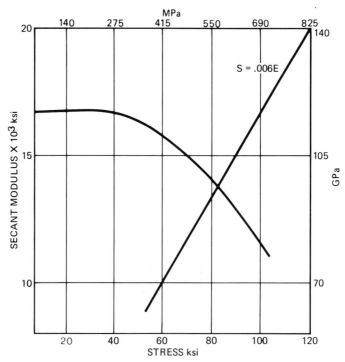

FIGURE 13–13
Graphical solution for stress in a cantilever beam.

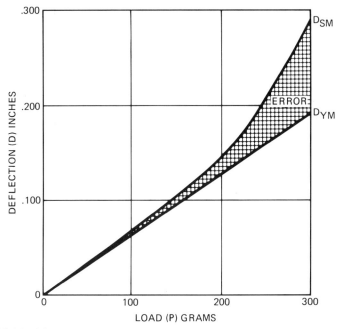

FIGURE 13–14
Error introduced in calculating deflection at various loads when modulus is assumed to be constant.

In order to more accurately predict behavior, the secant modulus is again used in place of Young's modulus. But the secant modulus is a function of the stress! Without knowing the stress (which is the calculation desired), the correct modulus value cannot be selected from Figure 13–12. This dilemma can be resolved by using reiterative calculation:

$$S = 0.006E, E = E_y = 16.8 \times 10^6, \quad S = 101$$

$$\text{At } S = 101, E_{sm} = 11.5 (E_{sm} = \text{secant modulus at a stress of 101 ksi} \\ \text{from Figure 13–12})$$

$$\text{At } E_{sm} = 11.5, S = 69 \text{ (from equation)}$$

Repeating this process (obviously a minicomputer can be programmed to facilitate the calculation) gives

$$
\begin{array}{ll}
\text{At } S = 69, & E_{sm} = 14 \\
\text{At } E_{sm} = 14, & S = 84 \\
\text{At } S = 84, & E_{sm} = 13.2 \\
\text{At } E_{sm} = 13.2, & S = 79 \\
\text{At } S = 79, & E_{sm} = 13.8
\end{array}
$$

$$At\,E_{sm} = 13.8, \qquad S = 83$$
$$At\,S = 83, \qquad E_{sm} = 13.3$$
$$At\,E_{sm} = 13.3, \qquad S = 80$$
$$At\,S = 80, \qquad E_{sm} = 13.8$$

If this calculation is continued, the value for the secant modulus will reach 13.6×10^6 psi and the predicted stress is

$$S = 0.006E$$

$$S = 0.006 \times 13.6 \times 10^6$$

$$S = 82 \text{ ksi}$$

The original stress calculation of 101 ksi is quite inaccurate, as expected. Compared to the subsequent stress calculation of 82 ksi, the initial result contains 23% error!

Since the "correct" modulus and stress obviously lie on the curve relating stress and modulus and since the stress-modulus values must also satisfy the equation $S = 0.006E$, a graphical solution can be used rather than the tedius reiteration method.

Figure 13–13 is nothing more than Figure 13–12 with the above linear equation superimposed. The solution previously found by reiteration is at the intersection of the two lines, that is, stress equals 82 ksi, secant modulus equals 13.6×10^6 psi.

Adjustment for Stress Relaxation

Not only is the initial contact force of 250 g important, but so also is the contact force after a given time period in service. Assume the criterion is 230 g after 5 years. Will Spring temper Alloy C51000 satisfy this requirement?

Figure 13–15 is a representative relaxation curve for this material loaded longitudinally in flexure, with an initial stress (S_o) of 80 ksi. After 5 years the stress remaining is approximately 77 ksi or 96% of the original stress.

Since contact force is directly proportional to stress, the contact force after 5 years is $P_F = 0.96 \times 250 = 240g$
Thus the requirement is satisfied.

Another technique for solving this problem is to "back" into a material. In this case, the designer would begin by calculating that, if the final force must be 230 g remaining after 5 years and the initial force is 250 g, the minimum percent of stress remaining is

$$\frac{230 \times 100}{250} = 92\%$$

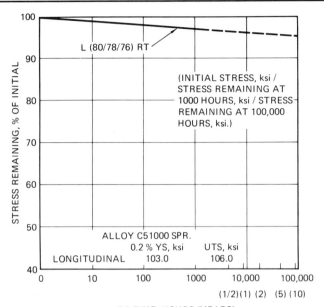

FIGURE 13-15
Stress relaxation of Alloy C51000 in flexure, longitudinal direction.

The designer can then search the relaxation curves for a material that meets this criterion. Although several materials exceed this requirement, for simplicity, assume the curve for Spring temper Alloy C51000 is selected. This curve shows that the actual stress remaining is 96% after 5 years. If the final force is 230 g (0.506 lb), the stress required to produce this force can be determined from Figure 13-10, equation (1), to be

$$S = \frac{6 \times 0.506 \times 1}{0.1 \times 0.02^2} = 76,000 \text{ psi}$$

The original force which would have been present before 5 years of relaxation is

$$0.506 \times \frac{1}{0.96} = 0.527 \text{ lbf}$$

Similarly, the initial stress would be

$$76,000 \times \frac{1}{0.96} = 79,200 \text{ psi}$$

With these values established, the designer can now determine the initial deflection. Knowing the mode, orientation, and stress, he can select the appropriate curve representing Alloy C51000 (Figure 13–12) and read off the value for secant modulus at the calculated stress. In this case, $E = 13.6 \times 10^6$ psi. This technique circumvents the need for either the graphical or reiterative solution for the modulus.

The deflection is calculated from equation (2) of Figure 13–10, using the above force and secant modulus values:

$$\frac{4 \times 1^3 \times 0.527}{0.1 \times 0.02^3 \times 13.6 \times 10^6} = 0.194 \text{ in.}$$

With alloy, temper, mode, orientation, and deflection established, the designer may now develop the final part layout.

Restrictions

A specifically dimensioned cantilever design was used throughout this analysis, but any terminal or connector which is basically a cantilever can be evaluated with this technique by making appropriate changes in geometry. But the graph of secant modulus versus stress is valid for only the particular alloy and temper shown, and only if the bending axis is transverse to the rolling direction. The Olin Brass Metals Research Laboratory is currently developing data for other alloys, tempers, and so on. This work has already shown that, while Young's modulus is almost unaffected by mill processing of sheet and strip, the secant modulus can be significantly altered, and in a controlled way. Thus the sophisticated designer will have another "tool" that can be utilized in effecting optimization.

The foregoing methods of calculation involve some approximations, and actual experience shows that predicted values are not always exact. However, they are much more accurate than the results obtained when the classical methods are used. Model shop and prototype work cannot be eliminated, but designers will find such work greatly facilitated because of the improved accuracy.

TESTING

Once the material selection is made and sample parts fabricated, they will need to be tested to insure they meet the design criteria. The importance of the testing procedure is obvious: too strenuous tests could result in an overdesigned, excessively expensive part, whereas tests not strenuous enough could eventually result in field failures.

Temperature is one factor in test procedures that is easily misapplied. Because of economic and time constraints it is seldom that a spring can be tested over the entire period of its expected life. In an attempt to circumvent this situation, time to failure can be accelerated by raising the temperature, assuming that this is equivalent to the effect of longer times at normal service temperature.

The acceleration to failure caused by elevated temperature is normally related to the Arrhenius equation, which states that the speed of chemical reactions varies with temperature as follows:

Rate of Process $= K \exp - \dfrac{Q}{RT}$ where K and R are constants, Q is the

activation energy of the process, and T is the temperature in degrees Celcius.

For the copper alloys, the value of Q is not always a constant, but changes at relatively low temperatures, as shown in Figure 13–16. Also the temperature and the magnitude of the change vary from alloy to alloy. Hence the accelerating effect is not always proportional to the increase in temperature nor is it consistent from one alloy to another, a fact which throws question on the validity of the results.

Figure 13–17 shows comparative stress relaxation data for Alloys C51000 and C76200. At room temperature their relaxation rates are

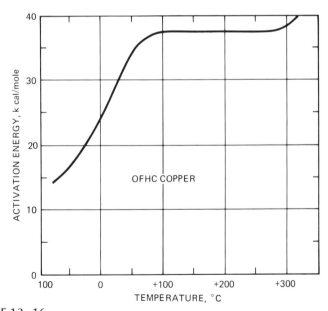

FIGURE 13–16
Variation of activation energy (Q) of oxygen-free, high conductivity copper with temperature. (Unpublished Olin Research Lab. Report.)

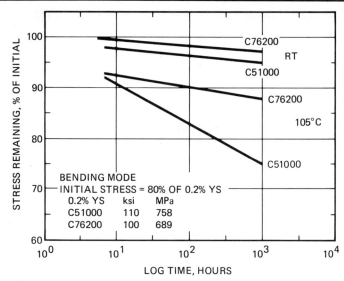

FIGURE 13–17
Effect of temperature on stress relaxation rates of Alloys C51000 and C76200.

very similar; however at 105°C (221°F), there is substantial difference. Suppose a part is expected to operate at room temperature, but in order to accelerate the testing the designer decided to increase the temperature. In testing alloys C51000 would appear to be substantially inferior, although in actual room temperature service it would perform just as well as C76200.

In lieu of accelerated testing, extrapolation of short-time test data is recommended. Experience has shown that this technique is accurate if the testing is continued long enough to establish the basic response trend. In testing stress relaxation, Olin's Metals Research Laboratory has determined that 2000 hour data can be extrapolated to adequately predict 10 year performance.

Table 13–9 shows an instance that supports this extrapolation. The laboratory's extrapolation was based upon 2000 hour data, and at both room temperature and elevated temperature the results agree very closely with actual long-time testing.

FORMULAS

For the convenience of the reader, Figure 13–18 gives some basic mathematical formulas often used in spring design. For more complex geometries or loadings the reader is referred to an appropriate engineering text.

TABLE 13-9
Accuracy of Extrapolated Test Data for Alloy C51000

Test Temperature	Initial Stress		Stress Remaining after 100,000 Hours			
			By Extrapolation		Actual Long-Time Test	
	ksi	Mpa	ksi	Mpa	ksi	Mpa
Room	97	669	89	614	88	607
105C	95	655	58	400	62	427

Source—see Reference 4.

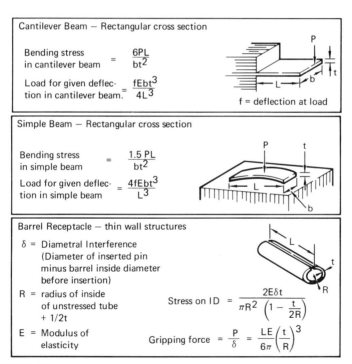

Cantilever Beam — Rectangular cross section

Bending stress in cantilever beam $= \dfrac{6PL}{bt^2}$

Load for given deflection in cantilever beam. $= \dfrac{fEbt^3}{4L^3}$

f = deflection at load

Simple Beam — Rectangular cross section

Bending stress in simple beam $= \dfrac{1.5\,PL}{bt^2}$

Load for given deflection in simple beam $= \dfrac{4fEbt^3}{L^3}$

Barrel Receptacle — thin wall structures

δ = Diametral Interference (Diameter of inserted pin minus barrel inside diameter before insertion)

R = radius of inside of unstressed tube + 1/2t

E = Modulus of elasticity

Stress on ID $= \dfrac{2E\delta t}{\pi R^2 \left(1 - \dfrac{t}{2R}\right)}$

Gripping force $= \dfrac{P}{\delta} = \dfrac{LE}{6\pi}\left(\dfrac{t}{R}\right)^3$

FIGURE 13–18
Basic formulas for spring design. Note that, when stresses exceed the elastic limit, *E* should be set equal to the secant modulus.

NEW DATA

Spring design information is available in the brochures of many of the companies that specialize in the manufacture of springs, such as Associated Spring Corporation, Instrument Specialties Company, and members of the Spring Manufacturers Institute. For some spring applications, however, particularly in small electrical and electronic assemblies in which copper alloys are used, new concepts and test data not so readily available have been needed by spring designers. To allow broader application of these more refined techniques, Olin Brass has published *The Spring Designer's Data Package.* This contains yield strength, fatigue, stress relaxation, and secant modulus data for a wide variety of copper alloys.

BIBLIOGRAPHY

1. Crane, J., "Performance and Fabricability Requirements for Copper-Base Alloys in Electrical and Electronic Connectors, "Society of Manufacturing Engineers, Dearborn, Mich, 1976.
2. Crane, J., P. Parikh, and E., Shapiro, "Mechanical Metallurgy of Copper-Base Alloy Electronic Connectors, "Olin Metals Research Laboratories, New Haven, Conn, 1974.
3. Pitney, K. E., *Ney Contact Manual*, The J. M. Ney Co., Bloomfield, Conn., 1973.
4. "Surface and Bulk Properties of Copper-Base Alloys for Electrical and Electronic Springs," Olin Metals Research Laboratories, New Haven, Conn., 1975, p. 15.
5. Zarlingo, S. P., and P., Parkh, "Predicting Contact Force, Stress, and Deflection in Connector Designs," *Ninth Annual Connector Symposium Proceedings*, Electronic Connector Study Group, Cherry Hill, N.J., 1976.
6. Zarlingo, S. P., "Predicting Contact Force, Deflection, and Stress in a Terminal/ Connector Design," Olin Brass, East Alton, Ill, 1976.

14

SPECIFYING
PRODUCT QUALITY

HOW QUALITY IS DESCRIBED

"Quality" is often used as though it were a definitive term. A product may be described as high quality or, in contrast, as low quality. Actually, "quality" used in this manner is ambiguous. "High quality" or "good quality" may mean different things to different people, such as "free from defects," or "produced from maximum-purity raw materials," or "produced by processes which have been controlled more stringently than required by standards." Finally, to some, it may mean "more thoroughly tested and inspected than standard products."

High or good quality copper and copper alloy sheet or strip should reflect all of these requirements and controls. Even more significantly, it should be manufactured to meet the particular demands of the part to be produced from it. This is the tailor-made or application-engineered feature, which in the final analysis is the real difference between good and poor quality. Quality basically reduces to how well a metal performs the task for which it was intended. If it can be fabricated without unusual difficulty and the finished part functions as intended, the metal has met the quality requirements admirably.

Looking at quality in these terms clearly suggests that a special relationship is desirable between the supplier and the user of copper alloy strip. Ready accessibility and open exchange of information between them are needed to assure that the specified quality of the material truly meets the requirements of the article to be fabricated from it. The user enters into the relationship with knowledge of the functional requirements of the part and with knowledge of how the part will be fabricated, finished, and assembled. The supplier has knowledge of the properties of the material, its responses in various fabricating situations, and the ways in which fabricating method can be adapted to solve anticipated problems. When a free and open exchange of this information takes place at both the engineering level and the sales-purchasing level, the most suitable quality and most appropriate cost are the usual result—and, finally, a useful article. Usefulness is defined here as *cost effectiveness*. It is the capability of the article to be fabricated and to perform the desired function at a cost acceptable to the manufacturer and consumer alike.

The preceding chapters contain information the manufacturer of copper and copper alloy sheet and strip can bring to the developmental meeting when assisting the user to choose an alloy with the appropriate qualities and minimum cost features for a particular application. The salesman and sales engineer, backed by metallurgical and research laboratories and process and manufacturing engineers, can be most helpful when fully informed about the manner in which the metal is to be used.

In the absence of a free exchange of information, material requirements are sometimes specified which contribute no value to the particular application. The metal manufacturer's costs are increased, and the end result is an inflated price. An example of this would be unnecessarily close thickness tolerances or camber tolerances or restrictive coil sizes or packaging.

The information given the supplier concerning the manner in which the metal will be used by the purchaser is the basis for the manufacturing specifications used in the supplier's plant. The marketing engineer, in consultation with process engineers and with the manufacturing and inspection supervisors, develops the manufacturing specifications. At the earliest stage in processing where subsequent operations will begin to influence the final specified properties and tolerances, metal in slab or coil form is taken from mill stock and applied to the order. The processing from this point on is then designed to develop the particular requirements of the user application.

All pertinent details of composition and other quality characteristics of each bar of metal are recorded from the time it is cast in the melt shop. These data are used to insure that the metal is applied where it is most suited. The inspection sequence is designed in parallel with the process, so all pertinent tests and inspections are made.

In the well-managed modern mill, production equipment is regularly monitored to insure that it is functioning correctly. Well-trained and informed production people know the quality requirements of each item and control the manufacturing operations so they are met. Since a poorly performed operation early in the process sequence will result in rejected product and poor yield in later steps, each operation is under careful control to provide a consistent-quality product.

The selection or preparation of specifications is an important and useful antecedent to the manufacture of a quality or cost-effective product. A first, essential step is to obtain a sample (if available), drawing, or sketch of the part to be fabricated from the strip. Information on the method of manufacture, assembly, and functioning of the part is most desirable for the engineer developing the metal specifications. In the case of a large or complex article, such as an automobile radiator, the most practical situation would be for the strip-supplier's representative to visit the manufacturer's plant to observe how the strip will be used. Plant visitations by a supplier's technical people are beneficial to any user. Direct observation of how the metal will be fabricated, and laboratory examination of the parts, are powerful tools for the development of effective specifications.

Different parts, for quite different end uses, often turn out to require essentially identical properites in the sheet and strip from which they will be manufactured. The same degree of corrosion resistance, similar tensile strength or hardness, and similar ductility may all be needed to satisfy the respective applications. For example, there are many different kinds of parts which are best made from 1/2 Hard temper Alloy C26000.

This commonality of requirments has led to the development and publication of standards and specifications by various specification-writing bodies. Manufacturers' associations have frequently written specifications covering the sheet and strip used by their members. The Society of Automotive Engineers' (SAE) specifications covering materials used in automobile parts are typical examples. The Copper Development Associaton publishes industry standards for copper and copper alloys.

The American Society for Testing and Materials (ASTM) is the organization responsible for the specifications most frequently used by the public as purchase specifications for copper and copper alloy sheet and strip. The ASTM uses a cooperative and voluntary standardization system. Its specification-writing committees consist of a balanced group of producers, consumers, and general interest members. They are brought together to write specifications, test methods, and recommended practices, and to exchange technical information. A consensus must be reached before a specification is published.

There are other national and even international organizations for the purpose of publishing standards and specifications that form the basis of trade and describe the basic requirements for many of the materials and equipment of the world's modern technology. The American National Standards Institute (ANSI) and the American Society of Mechanical

Engineers (ASME) publish standards based on ASTM specifications or specifications written by other standards-writing organizations. The ASME specifications may include design criteria for materials used in pressure vessels. The International Organization for Standardization (ISO) is the major international standards-writing organization in the world. Its standards-writing committees are made up of delegates who are technical experts and who represent the producer and consumer interests of their countries. ISO standards must be approved by the world body before they are published.

Published ASTM specifications contain all the basic requirements needed for the large majority of applications. Usually, ASTM specifications cover a family of related alloys. Thus ASTM B36 covers the copper-zinc alloys (commonly called brasses); ASTM B103, the copper-tin alloys (bronzes and phosphor bronzes); ASTM B121, the copper-zinc-lead alloys (leaded brasses); ASTM B152, the coppers; ASTM B122, the copper-nickel-zinc and copper-nickel alloys (nickel silvers and cupro-nickels); ASTM B591, the copper-zinc-tin alloys. Each of these specifications contains the following requirements and information:

1. A scope.
2. Ordering information.
3. Chemical compositions.
4. Tempers.
5. Grain size.
6. Tensile strength.
7. Hardness.

Each specification includes reference to ASTM B248, which adds the following requirements and information:

1. Product definitions and descriptions.
2. Applicable methods-of-test documents.
3. Thickness tolerances.
4. Width tolerances.
5. Length tolerances for lengths.
6. Camber tolerances.
7. Edge requirements.
8. Workmanship and finish.
9. Sampling.
10. Number of tests and retests.
11. Methods of test.
12. Significance of numerical limits.

When the end use demands some special property not included in the standard requirements, it can be added to the purchase order as a supplement. This procedure is followed by some of the large users of copper and copper alloy strip, who use ASTM specifications and add other special requirements. For example, coil size and packaging requirements designed to suit the handling equipment at a particular plant would be included.

DIMENSIONAL TOLERANCES FOR THICKNESS, WIDTH, LENGTH, CAMBER, AND FLATNESS

In the selection of an alloy for a particular application, basic physical properties such as electrical conductivity and corrosion resistance are the first factors to be considered. Variable qualities such as strength, thickness, and formability then may become trade-off situations. As an example, if the strength requirement can be met only by using an alloy of lower conductivity, a larger cross-sectional area may be needed to carry the electric current without excessive heat rise. Cost, too, is a major factor in choice of material.

Once the basic selection has been made, dimensional tolerances can be established, based on such considerations as the part fabricating equipment, fit-up required in assembly, and functional requirements of the finished part.

THICKNESS TOLERANCE

Thickness tolerance is the allowable variation from the specified gauge or thickness of the strip or sheet. The extent of such variation primarily depends on the capabilities of the rolling mill equipment. Every mill operation which reduces the thickness of the product potentially can introduce thickness variations from end to end or from edge to edge of a coil of metal being processed. The thickness tolerances given in Tables 14–1, 14–2, and 14–3 from ASTM B248[1] for refractory and nonrefractory alloys are widely used and acceptable standards which represent the experience of brass mill production over a long period of time. If no thickness tolerance is specified on a purchase order, these are applied.

The thickness tolerances for *refractory* alloys are greater than those for *nonrefractory* alloys because refractory alloys are more resistant to deformation and are difficult to work. Therefore, greater potential for variation exists.

Modern rolling mills and sophisticated gauging devices have contributed to the mills' capability for controlling thickness variations. Con-

[1] Standard Specification for General Requirements for Wrought Copper and Copper Alloy Plate, Sheet, Strip and Rolled Bar.

TABLE 14-1
Thickness Tolerances for Nonrefractory Alloys[a]
(Applicable to Specifications B36, B121, B152, and B465)

Thickness Tolerances, plus and minus,[b] in.

Thickness, in.	8 in. and Under in Width	Over 8 to 12 In., Incl., in Width	Over 12 to 14 In., Incl., in Width	Over 14 to 20 In., Incl., in Width	Over 20 to 28 In., Incl., in Width	Over 28 to 36 In., Incl., in Width	Over 36 to 48 In., Incl., in Width	Over 48 to 60 In., Incl., in Width
	Strip			**Sheet**				
0.004 and under	0.0003	0.0006	0.0006	⋯	⋯	⋯	⋯	⋯
Over 0.004 to 0.006, incl.	0.0004	0.0008	0.0008	0.0013	⋯	⋯	⋯	⋯
Over 0.006 to 0.009, incl.	0.0006	0.0010	0.0010	0.0015	⋯	⋯	⋯	⋯
Over 0.009 to 0.013, incl.	0.0008	0.0013	0.0013	0.0018	0.0025	0.003	0.0035	0.004
Over 0.013 to 0.017, incl.	0.0010	0.0015	0.0015	0.002	0.0025	0.003	0.0035	0.0045
Over 0.017 to 0.021, incl.	0.0013	0.0018	0.0018	0.002	0.003	0.0035	0.004	0.005
Over 0.021 to 0.026, incl.	0.0015	0.002	0.002	0.0025	0.003	0.0035	0.004	0.005
Over 0.026 to 0.037, incl.	0.002	0.002	0.002	0.0025	0.0035	0.004	0.005	0.006
Over 0.037 to 0.050, incl.	0.002	0.0025	0.0025	0.003	0.004	0.005	0.006	0.007
Over 0.050 to 0.073, incl.	0.0025	0.003	0.003	0.0035	0.005	0.006	0.007	0.008
Over 0.073 to 0.130, incl.	0.003	0.0035	0.0035	0.004	0.006	0.007	0.008	0.010
Over 0.130 to 0.188, incl.	0.0035	0.004	0.004	0.0045	0.007	0.008	0.010	0.012
	Rolled Bar			**Plate**				
Over 0.188 to 0.205, incl.	0.0035	0.004	0.004	0.0045	0.007	0.008	0.010	0.012
Over 0.205 to 0.300, incl.	0.004	0.0045	0.0045	0.005	0.009	0.010	0.012	0.014
Over 0.300 to 0.500, incl.	0.0045	0.005	0.005	0.006	0.012	0.013	0.015	0.018
Over 0.500 to 0.750, incl.	0.0055	0.007	0.007	0.009	0.015	0.017	0.019	0.023
Over 0.750 to 1.00, incl.	0.007	0.009	0.009	0.011	0.018	0.021	0.024	0.029
Over 1.00 to 1.50, incl.	0.022	0.022	0.022	0.022	0.022	0.025	0.029	0.036
Over 1.50 to 2.00, incl.	0.026	0.026	0.026	0.026	0.026	0.030	0.036	0.044

[a] See Table 14-1A for metric conversion values.   [b] When tolerances are specified as all plus or all minus, double the values given.

TABLE 14-1A
Conversion Values for Tables 14-1 and 14-2

in.	mm	in.	mm	in.	mm	in.	mm	in.	mm
Width of Material									
8	203	14	356	28	711	48	1220		
12	305	20	508	36	914	60	1520		
Thickness of Material									
0.004	0.102	0.021	0.533	0.130	3.30	0.750	19.1		
0.006	0.152	0.026	0.660	0.188	4.78	1.00	25.4		
0.009	0.229	0.037	0.940	0.205	5.21	1.50	38.1		
0.013	0.330	0.050	1.27	0.300	7.62	2.00	50.8		
0.017	0.432	0.073	1.85	0.500	12.7				
Thickness Tolerances									
0.0003	0.0076	0.003	0.076	0.010	0.25	0.021	0.53	0.033	0.84
0.0004	0.010	0.0035	0.089	0.011	0.28	0.022	0.56	0.036	0.91
0.0006	0.015	0.004	0.10	0.012	0.30	0.023	0.58	0.037	0.94
0.0008	0.020	0.0045	0.11	0.013	0.33	0.024	0.61	0.038	0.97
0.0010	0.025	0.005	0.13	0.014	0.36	0.025	0.64	0.044	1.1
0.0013	0.033	0.0055	0.14	0.015	0.38	0.026	0.66	0.045	1.1
0.0015	0.038	0.006	0.15	0.016	0.41	0.028	0.71	0.055	1.4
0.0018	0.046	0.007	0.18	0.017	0.43	0.029	0.74		
0.002	0.051	0.008	0.20	0.018	0.46	0.030	0.76		
0.0025	0.064	0.009	0.23	0.019	0.48	0.032	0.81		

TABLE 14-2
Thickness Tolerances for Refractory Alloys[a]
(Applicable to Specifications B97, B103, B122, B169, B194, B422, and B534)

Thickness Tolerances, plus and minus,[b] in.

Thickness, in.	Strip — 8 In. and Under in Width	Strip — Over 8 to 12 In., Incl., in Width	Strip — Over 12 to 14 In., Incl., in Width	Strip — Over 14 to 20 In., Incl., in Width	Sheet — Over 20 to 28 In., Incl., in Width	Sheet — Over 28 to 36 In., Incl., in Width	Sheet — Over 36 to 48 In., Incl., in Width	Sheet — Over 48 to 60 In., Incl., in Width
0.004 and under	0.0004	0.0008	0.0008	…	…	…	…	…
Over 0.004 to 0.006, incl.	0.0006	0.0010	0.0010	0.0015	…	…	…	…
Over 0.006 to 0.009, incl.	0.0008	0.0013	0.0013	0.002	…	…	…	…
Over 0.009 to 0.013, incl.	0.0010	0.0015	0.0015	0.0025	…	…	…	…
Over 0.013 to 0.017, incl.	0.0013	0.002	0.002	0.0025	…	…	…	…
Over 0.017 to 0.021, incl.	0.0015	0.0025	0.0025	0.003	…	…	…	…
Over 0.021 to 0.026, incl.	0.002	0.0025	0.0025	0.003	0.004	0.005	0.006	0.007
Over 0.026 to 0.037, incl.	0.0025	0.003	0.003	0.0035	0.005	0.006	0.007	0.008
Over 0.037 to 0.050, incl.	0.003	0.0035	0.0035	0.004	0.006	0.007	0.008	0.010
Over 0.050 to 0.073, incl.	0.0035	0.004	0.004	0.0045	0.007	0.008	0.010	0.012
Over 0.073 to 0.130, incl.	0.004	0.0045	0.0045	0.005	0.008	0.010	0.012	0.014
Over 0.130 to 0.188, incl.	0.0045	0.005	0.005	0.006	0.010	0.012	0.014	0.016
	Rolled Bar				*Plate*			
Over 0.188 to 0.205, incl.	0.0045	0.005	0.005	0.006	0.010	0.012	0.014	0.016
Over 0.205 to 0.300, incl.	0.005	0.006	0.006	0.007	0.012	0.014	0.016	0.018
Over 0.300 to 0.500, incl.	0.006	0.007	0.007	0.008	0.015	0.017	0.019	0.023
Over 0.500 to 0.750, incl.	0.008	0.010	0.010	0.012	0.019	0.021	0.024	0.029
Over 0.750 to 1.00, incl.	0.010	0.012	0.012	0.015	0.023	0.026	0.030	0.037
Over 1.00 to 1.50, incl.	0.028	0.028	0.028	0.028	0.028	0.032	0.037	0.045
Over 1.50 to 2.00, incl.	0.033	0.033	0.033	0.033	0.033	0.038	0.045	0.055

[a]See Table 14-1A for metric conversion values.   [b]When tolerances are specified as all plus or all minus, double the values given.

TABLE 14-3
Special Thickness Tolerances

Thickness, in. (mm)	Tolerances Applicable to Alloy 725, Specification B 122 Tolerances, plus and minus, [a]in. (mm) for Strip 8 in. and Under in Width	Tolerances Applicable to Specifications B 194 and B 534 Tolerances, plus and minus, [a]in. (mm) for Strip 4 in. and Under in Width
0.004 (0.102) and under	0.0002 (0.005)	0.0002 (0.005)
Over 0.004 to 0.006, (0.102 to 0.152), incl.	0.0003 (0.008)	0.0003 (0.008)
Over 0.006 to 0.009, (0.152 to 0.229), incl.	0.0004 (0.010)	0.0005 (0.013)
Over 0.009 to 0.013, (0.229 to 0.330), incl.	0.0005 (0.013)	0.0006 (0.015)
Over 0.013 to 0.017, (0.330 to 0.432), incl.	0.0007 (0.018)	0.0007 (0.018)
Over 0.017 to 0.021, (0.432 to 0.533), incl.	0.0008 (0.020)	0.0008 (0.020)
Over 0.021 to 0.026, (0.533 to 0.660), incl.	0.0010 (0.025)	0.0010 (0.025)
Over 0.026 to 0.032, (0.660 to 0.813), incl.	0.0013 (0.033)	0.0010 (0.025)
Over 0.032 to 0.050, (0.813 to 1.27), incl.	0.0015 (0.038)	

[a]If tolerances are specified as all plus or all minus, double the values given.

versely, the width of metal being processed has increased, rolling mill speeds and applied forces have increased, and the length of the coils has increased, all to effect greater economy. These three factors make thickness control more difficult. When required, copper and copper alloy strip can be produced, by the use of extra or more time consuming operations, to meet more restrictive thickness tolerances. Such special requirements may be specified on the user's purchase order.

There is a limit to the tolerance that can be held for any given alloy, thickness, and temper. For example, when cold-rolling annealed metal to produce 1/4 Hard or 1/2 Hard temper, the very small reduction needed during this last operation makes control of the variations in thickness difficult.

Some alloys can be annealed to give the same tensile strength as produced by cold rolling. The user, in such cases, may find it an advantage to specify annealed-to-temper 1/4 Hard or 1/2 Hard for improved thickness uniformity as well as better formability. Since a greater reduction in thickness is used prior to annealing this product to temper, closer thickness control is possible.

WIDTH TOLERANCE

Width variations present few means of control to values closer than the standard requirements shown in Tables 14–4, 14–5, and 14–6. The majority of sheet and strip is cut to width on *rotary knife slitters*. The clearance between the matching knives required to produce a good edge with the least burr is governed largely by the thickness and temper of the alloy. Very thin metal requires little clearance; hence width variation between knives can be minimized, but the metal is also more easily distorted. With thicker metal, where a clearance between knives of 10% or more of the thickness may be needed, there is greater chance for width variation.

Each slitter-knife set is made up of a combination of knives, fillers, and shims. There is a set for each different width processed in the brass mill. All the strips sheared by any particular knife set will be quite uniform in width. However, when the same kind of material is to be slit to the same width at a future date, the slitter-knife set will have to be reassembled. The width will still be within the standard tolerance, but may be slightly different from the previous order.

This situation cannot be practically manipulated to give closer control of width. Therefore the standards given in ASTM Specification B248 generally represent the narrowest feasible width tolerances and are normally applied to all orders. Closer width tolerances than these may occasionally be possible on some items, but extra operations, such as edge rolling, may be necessary.

TABLE 14-4
Width Tolerances for Slit Metal and Slit Metal with Rolled Edges

Width, in. (mm)	Width Tolerances,[a] plus and minus							
	For Thicknesses 0.004 to 0.032 In. (0.102 to 0.813 mm), Incl.		For Thicknesses over 0.032 to 0.125 In. (0.813 to 3.18 mm), Incl.		For Thicknesses over 0.125 to 0.188 In. (3.18 to 4.78 mm), Incl.		For Thicknesses over 0.188 to 0.500 In. (4.78 to 12.7 mm), Incl.	
	in.	mm.	in.	mm.	in.	mm.	in.	mm.
2 (50.8) and under	0.005	0.13	0.010	0.25	0.012	0.30	0.015	0.38
Over 2 to 8 (50.8 to 203), incl.	0.008	0.20	0.013	0.33	0.015	0.38	0.015	0.38
Over 8 to 24 (203 to 610), incl.	0.015	0.40	0.015	0.40	0.015	0.40	0.031	0.79
Over 24 to 40 (610 to 1020), incl.	0.031	0.79	0.031	0.79	0.031	0.79	0.046	1.19

[a]If tolerances are specified as all plus or all minus, double the values given.

TABLE 14-5
Width Tolerances for Square-Sheared Metal
All Lengths up to 120 In. (3.05 m), Incl.

	Width Tolerances,[a] plus and minus					
	$\frac{1}{16}$ In. (1.59 mm) and Under in Thickness		Over $\frac{1}{16}$ (1.59 mm) to $\frac{1}{8}$ In. (3.18 mm), Incl., in Thickness		Over $\frac{1}{8}$ In. (3.18 mm) in Thickness	
Width, in. (mm)	in.	mm	in.	mm	in.	mm
20 (508) and under	$\frac{1}{32}$	0.79	$\frac{1}{64}$	1.2	$\frac{1}{16}$	1.6
Over 20 (508) to 36 (914), incl.	$\frac{1}{64}$	1.2	$\frac{1}{64}$	1.2	$\frac{1}{16}$	1.6
Over 36 (914) to 120 (3050), incl.	$\frac{1}{16}$	1.6	$\frac{1}{16}$	1.6	$\frac{1}{16}$	1.6

[a]If tolerances are specified as all plus or all minus, double the values given.

TABLE 14-6
Width Tolerances for Sawed Metal

	Width Tolerances,[a] plus and minus					
	For Lengths up to 10 Ft (305 m), Incl.				For Length over 10 Ft (3.05 m)	
	For Thicknesses up to 1½ In. (38.1 mm), Incl.		For Thicknesses over 1½ In. (38.1 mm)		All thicknesses	
Width, in. (mm)	in.	mm	in.	mm	in.	mm
Up to 12 (305), incl.	$\frac{1}{32}$	0.79	$\frac{1}{16}$	1.6	$\frac{1}{16}$	1.6
Over 12 (305) to 120 (3050), incl.	$\frac{1}{16}$	1.6	$\frac{1}{16}$	1.6	$\frac{1}{16}$	1.6

[a]If tolerances are specified as all plus or all minus, double the values given.

LENGTH TOLERANCE

Length tolerances for product straightened and then cut to length are also given in ASTM B248. Most lengths are *automatically machine cut*. The tolerances are determined by the machine capabilities. When closer tolerances are required, the lengths must be individually recut by hand on a reshear. Table 14–7 gives the tolerances for lengths which are automatically machine cut. This is the common method, and these tolerances apply to all cut lengths unless special cutting methods are specified. Table 14–9 lists the length tolerances which apply to square-sheared lengths, which are cut on manually operated equipment to provide restricted tolerances and good, square cut ends. Table 14–10 gives the length tolerances which apply to metal that is sawed to length. In general, sawing is applied only to metal which is too thick for shear cutting.

When long coils are flattened and cut to lengths, shorter-than-nominal end pieces are often allowed. The length and the number of such pieces allowed are limited to the schedule given in Table 14–8.

CAMBER OR STRAIGHTNESS TOLERANCE

Camber, or edgewise curvature, requires a different method of describing tolerance. The standard in this case is based on the departure of the strip edge from a straight line of stated length. The consistent production of slit edge strip which is absolutely straight is not possible, and some allowance for variation from straightness must be recognized. Straightness tolerances for sheet and strip are given in Tables 14–11, 14–12, and 14–13, taken from ASTM B248. When no camber or straightness tolerance is specified by the purchaser of sheet or strip, these tolerances are applied by the producer.

The control of straightness involves almost every part of the strip manufacturing operation. To produce the straightest possible strip, it is imperative that the coil of metal arrive at the slitting operation reasonably straight and free from excessive waviness along the edges or

TABLE 14-7
Length Tolerances for Straight Lengths

Length, ft (m)	Length Tolerances[a]	
	in.	mm
Specific lengths, mill lengths, multiple lengths, and specific lengths with ends		
10 (3.05) and under	$\frac{1}{4}$	6.4
Over 10 (3.05) to 20 (6.10), incl.	$\frac{1}{2}$	13
Stock lengths and stock lengths with ends	1^b	25^b

[a]The length tolerances are all plus; if all minus tolerances are desired, use the same values; if plus and minus tolerances are desired, halve the values given.
[b]As stock lengths are cut and placed in stock in advance of orders, departure from this tolerance is not practicable.

TABLE 14-8
Schedule of Minimum Length and Maximum Weight of Ends for Mill Lengths, Specific Lengths with Ends, and Stock Lengths with Ends

Nominal Length, ft (m)	0.050 In. (1.27 mm) and Under in Thickness			Over 0.050 to 0.125 In. (1.27 to 3.18 mm), Incl., in Thickness			Over 0.125 to 0.250 In. (3.18 to 6.35 mm), Incl., in Thickness		
	Minimum Length of Shortest Piece		Maximum Permissible Weight of Ends, % of lot weight	Minimum Length of Shortest Piece		Maximum Permissible Weight of Ends, % of lot weight	Minimum Length of Shortest Piece		Maximum Permissible Weight of Ends, % of lot weight
	ft	m		ft	m		ft	m	
6 (1.83) to 8 (2.44), incl.	4	1.22	20	4	1.22	25	3	0.914	30
8 (2.44) to 10 (3.05), incl.	6	1.83	25	5	1.52	30	4	1.22	35
10 (3.05) to 14 (4.27), incl.	7	2.13	30	6	1.83	35	5	1.52	40

TABLE 14-9

Length Tolerances for Square-Sheared Metal in All Widths 120 In. (3.05 m) and Under

	Length Tolerances,[a] plus and minus					
	For Thickness up to $\frac{1}{16}$ In. (1.59 mm), Incl.		For Thicknesses over $\frac{1}{16}$ to $\frac{1}{8}$ In. (1.59 to 3.18 mm), Incl.		For Thicknesses over $\frac{1}{8}$ In. (3.18 mm)	
Length, in. (m)	in.	mm	in.	mm	in.	mm
20 (0.508) and under	$\frac{1}{32}$	0.79	$\frac{3}{64}$	1.2	$\frac{1}{16}$	1.6
Over 20 to 36 (0.508 to 0.914), incl.	$\frac{3}{64}$	1.2	$\frac{3}{64}$	1.2	$\frac{1}{16}$	1.6
Over 36 to 120 (0.914 to 3.05), incl.	$\frac{1}{16}$	1.6	$\frac{1}{16}$	1.6	$\frac{1}{16}$	1.6

[a]If tolerances are specified as all plus or all minus, double the values given.

TABLE 14-10

Length Tolerances for Sawed Metal

	Length Tolerance[a]	
Width, in. (mm)	in.	mm
Up to 120 (3050), incl.	$\frac{1}{4}$	6.4

[a]The tolerances are all plus: if all minus tolerances are desired, use the same values; if plus and minus tolerances are desired, halve the values given.

from buckles or fullness of shape. These conditions usually are the result of lack of thickness uniformity from edge to edge which developed at one of the early rolling operations. The edges of wide coils tend to be thinner than the center because the edge is not confined during rolling. Such an edge is trimmed and discarded at the final slitting operation, but strips nearest to it will tend to have more curvature than others slit from the same coil. When narrow widths come from wide process coils, they are sometimes fanned out for rewinding and some curvature can be induced. This is less of a problem with modern slitters equipped with tension devices and overarm separators. Nevertheless, it is sometimes necessary to split a coil at an earlier process stage, prior to final slitting, to achieve maximum straightness on narrow widths.

TABLE 14-11
Straightness Tolerances for Slit Metal or Slit Metal Either Straightened or Edge-Rolled
Maximum Edgewise Curvature (Depth of Arc) in any 72-In. (1.83 m) Portion of the Total Length

Width, in. (mm)	Straightness Tolerances					
	As Slit Only				As Slit and Either Straightened or Edge-Rolled	
	Shipped in Rolls		Shipped Flat		Shipped Flat, in Rolls, or on Bucks	
	in.	mm	in.	mm	in.	mm
Over $\frac{1}{4}$ to $\frac{3}{8}$ (6.35 to 9.53), incl.	2	51	$1\frac{1}{2}$	38	$\frac{1}{2}$	13
Over $\frac{3}{8}$ to $\frac{1}{2}$ (9.53 to 12.7), incl.	$1\frac{1}{2}$	38	1	25	$\frac{1}{2}$	13
Over $\frac{1}{2}$ to 1 (12.7 to 25.4), incl.	1	25	$\frac{3}{4}$	19	$\frac{1}{2}$	13
Over 1 to 2 (25.4 to 50.8), incl.	$\frac{5}{8}$	16	$\frac{5}{8}$	16	$\frac{3}{8}$	9.5
Over 2 to 4 (50.8 to 102), incl.	$\frac{1}{2}$	13	$\frac{1}{2}$	13	$\frac{3}{8}$	9.5
Over 4 (102)	$\frac{3}{8}$	9.5	$\frac{3}{8}$	9.5	$\frac{3}{8}$	9.5

TABLE 14-12
Straightness Tolerances for Square-Sheared Metal
Maximum Edgewise Curvature (Depth of Arc) in any 72 In. (1.83 m) Portion of the Total Length.
Not applicable to metal over 120 in. (3.05 m) in length.

| Thickness, in. (mm) | Straightness Tolerances | | | |
| | Up to 10 In. (254 mm), Incl., in Width | | Over 10 In. (254 mm), in Width | |
	in.	mm	in.	mm
$\frac{1}{8}$ (3.18) and under	$\frac{1}{16}$	1.6	$\frac{1}{32}$	0.79
Over $\frac{1}{8}$ to $\frac{3}{16}$ (3.18 to 4.78), incl.	$\frac{1}{8}$	3.2	$\frac{3}{64}$	1.2
Over $\frac{3}{16}$ (4.78)	$\frac{1}{8}$	3.2	$\frac{1}{16}$	1.6

TABLE 14-13
Straightness Tolerances for Sawed Metal
Maximum Edgewise Curvature (Depth of Arc) in any 72 In. (1.83 m) Portion of the
Total Length.
Not applicable to metal over 144 in. (3.66 m) in length.

Width, in. (mm)	Straightness Tolerances	
	in.	mm
3 (76.2) and under	$\frac{1}{16}$	1.6
Over 3 (76.2)	$\frac{3}{64}$	1.2

By carefully controlling rolling practices, camber can be held within the tolerance for most sizes. When straightness beyond what is possible in slitting practice is required, it becomes necessary to put the individual strips through added straightening or reroll operations. This is a slow, high-cost operation.

FLATNESS TOLERANCE

As camber is the measured departure from a straight line, flatness deviations are measured as the departure from a horizontal flat surface. All sheet and strip is processed to be flat. As there are no published standards covering flatness of copper alloy sheet and strip, these requirements are essentially qualitative. When coiled sheet or strip is required to be as flat as possible, it is customary to specify "free from waves and buckles."

When straight lengths are produced, the coiled metal is passed through a multiroll flattener before being cut, and relatively flat metal is produced. When extremely flat metal is required and the alloy size and temper are amenable, some cut lengths can be reflattened by *stretcher* or *patent leveling*.

There are other special requirements which relate to the dimensions, such as squareness of cut for straight lengths or coil set or curl in coiled metal, but there are no established standards. When a problem with such a characteristic is experienced or anticipated, the user should negotiate with the supplier to establish a mutually acceptable limit and a method of measurement. When coiled metal is to be used and the size and temper are such that inherent coil set is going to be present, then the only way to be sure of flatness control is for the user to equip his fabricating line with a multiroll machine capable of flattening the metal as it is fed into the blanking operation.

OTHER QUALITY FEATURES, DEFECTS, AND DIFFICULTIES

There are four important basic characteristics of a metal which influence its formability in press tools:

1. Alloy or composition (different alloys have different degrees of formability).
2. Temper (different tempers of the same alloy have different degrees of formability).
3. Gauge or thickness (the thickness of the metal to be formed must be taken into account in setting up forming tools).
4. Surface quality (the frictional forces between the tools and metal being formed must be under control).

The influences of alloy, temper, and thickness on formability are quite generally recognized, but not infrequently the influence of surface conditions is overlooked or misunderstood. When considering the influence of surface on formability, the type of lubrication which will be used must be known. In Chapter 2, "Melting and Mill Processing Operations," the types of surface conditions which may be present were described. In Chapter 10, "Fabrication by Forming, Drawing, and Related Operations," various types of lubricants were described. It is not the purpose at this point to repeat this information, but to consider surface in terms of specifications or lack of them.

A variety of surface qualities can result from the different processes or equipment which can be used to produce the same specified alloy, temper, and gauge. Each different brass mill produces surfaces which are typical of its particular equipment and processing techniques. Therefore there are no published standard specifications covering surface, except in general terms. Brass mills have in-house surface specifications, and when the customer informs the mill how the metal is to be used, the appropriate in-house specification will be applied and met and the metal should be suitable for the intended use.

Depending upon the type of mill and the lubricants used while rolling, metal may have anything from a dull, matte surface, at one extreme, to a shiny, burnished surface at the other. When forming parts in press tools from metal with such differences in surface, a thin lubricant-water solution may be suitable on the burnished metal, but insufficient to protect the matte surface from being scored and the tools from loading. A thicker, higher-film-strength lubricant, however, will protect both metal and tools regardless of the degree of burnish or matte finish. A supplier of strip who knows that a user's lubricant choice favors a burnished surface will try to provide it. It should be realized, however, that the roll lubricants and techniques which must be used by the mill to give burnished finishes tend also to produce streaked and stained surfaces.

Therefore the user who requires a burnished surface just to satisfy the lubricant situation in his plant courts other surface problems. What are normal variations in strip surface can create what may seem to be a major problem if the press lubricant itself is not adequate.

Surfaces of copper and copper alloy sheet and strip also vary, depending on the annealing practices and equipment. When soft, or annealed, temper is produced, for example, the surface may differ in color from the intrinsic color of the alloy. When coils are annealed in batch type furnaces, the exposure of the overlapping metal strips to the furnace atmosphere varies from edge to edge and end to end. The amount of surface variation depends on the annealing temperature required and the furnace atmosphere used. When a high zinc brass is annealed to develop a large grain size, some dezincification of the exposed edges is bound to occur. This can result, sometimes, in red stains after subsequent cleaning in a sulfuric acid solution. While modern annealing and cleaning practices minimize this discoloration, the portion of the surface where dezincification occurred will tend to have a slightly different, grayish brass color. In addition, the length of strip that can be annealed as coils in a batch type furnace is limited. The metal expands as the temperature is raised; and when the coil is large, the pressure between wraps is so great that one coil wrap welds to the adjacent one and the metal is severely damaged.

The vertical, high speed, continuous strip annealer has eliminated this problem as well as the surface variations which can occur when coiled metal is annealed. In this method the metal is uncoiled and degreased, annealed, and then cleaned in one continuous line. The resulting surface is uniform from edge to edge and from end to end.

The furnace atmosphere for continuous annealing is usually the product of combustion of the gas and air mixture fed to the burners used to heat the strip. With brass the fuel mixture must have a slight excess of air so the atmosphere will be somewhat oxidizing. This is necessary because the zinc in brass vaporizes at a low temperature and is given off as fumes during annealing. These zinc fumes attack furnace structures rapidly. Hence the atmosphere must contain enough oxygen to react with them to form zinc oxide and protect the furnace. The small excess of oxygen produces a thin oxide film on the strip surface. This is mostly removed in the cleaning tanks in the process line, but the trace of residual film remaining results in a characteristic color on the surface different from the natural color. In copper alloys containing zinc the residual film is usually quite beneficial because it is mostly zinc oxide, which protects the surface and aids lubrication in forming and deep-drawing operations.

Other coppers and copper alloys each have characteristic surfaces resulting from the annealing and cleaning practices used. Such differences do not affect the forming or deep-drawing operations if proper lubricants are used. Other than zinc oxide, most metal oxides, including copper oxide, have higher coefficients of friction with tools than the metal itself. Adequate lubrication is essential to avoid tool wear, part scoring, tool loading, and other surface related problems which can occur with oxidized metal surfaces.

Some copper alloys contain elements which produce extremely abrasive oxides. In the production of these alloys in strip form the amount of oxidation allowed in annealing must be carefully controlled, and the cleaning practice is designed to remove the refractory oxides. When such special care is not taken, rapid tool wear may be experienced. The tin bronzes or phosphor bronzes are prone to such harsh surfaces unless annealed under controlled conditions and protective atmospheres and cleaned with appropriate sulfuric acid solutions.

The cold-rolling, annealing, and cleaning processes are those which have the greatest influence on strip surfaces. *Buffability* (ability to polish and buff to a high luster) is a special surface characteristic. When good buffability is desired, the user should indicate this to the supplier, who will adjust each operation to provide the smoothest, cleanest metal possible. Certain pieces of equipment are best suited to produce buffable surfaces. Rolling mill rolls must be clean and smooth, and rolling practices must be carefully controlled to avoid damaging the metal surfaces.

Annealing practices are adjusted to provide the smallest grain size compatible with the temper required. Forming or drawing metal with a large grain size develops a characteristic roughening of the surface described by the term *orange peel*. When 1/4 Hard or Half Hard temper brass is to be formed into parts and then polished and buffed, it is recommended that 1/4 Hard annealed to temper of 1/2 Hard annealed to temper be used. These are made by fine-grain annealing processes and will avoid possible orange peel. When metal is being produced for buffability, special inspections and tests are added. Samples are taken at appropriate stages in processing and actually buffed, to be sure that the surfaces are suitable. Handling of the large coils of such special product must be mechanized and carefully supervised to avoid damaging or marring the surface.

Some surface defects which interfere with forming or buffing, as well as other operations such as soldering, brazing, or plating, can occur after the metal is shipped from the brass mill. All metal, as it is shipped, will contain a protective film of lubricant on the surfaces unless the mill is specifically requested to ship *dry*. Dry metal, however, is extremely prone to surface damage. Coils must be wound tightly or they will be easily damaged in transit. Tension must be maintained during winding, and some slippage between wraps always takes place. When the surfaces are dry, friction scratches occur. Further, during uncoiling in the user's plant, the release of tension again can cause friction scratches. If the metal user must have metal that is dry and free of such damage, the most practical approach is to arrange to remove the protective surface lubricant after shipment as part of the fabricating process. In this way, the user can be assured of having both dry and undamaged metal. In some instances the mill can interleave the metal with a fine paper, which affords some protection, but the user must have the proper paper handling facilities and recognize the cost involved.

A surface defect associated with how tightly the coils are wound and with the packaging and handling of the metal in transit is *fretting corrosion*. Fretting corrosion, or *black spots,* occur during transit when adjacent metal surfaces rub together from vibration or movements of the vehicle. The friction between adjacent surfaces causes minute particles to be gouged out, which oxidize rapidly and turn black. The surfaces are marred by a series of black pits at each point of intimate contact.

Coils of metal must be wound tightly at the mill to prevent movement between wraps as much as possible. A lubricant film provides some protection, but the coils still must be securely packaged to prevent shifting and vibration during transit. Some large coils can be packaged with the coil axis in the horizontal position, but for coils of such size that there is any tendency to sag, packing should be done with the axis in the vertical position. Properly coiled and packaged metal should reach the user in good condition unless the transporter has mishandled it. Any evidence of rough handling should be called to the attention of the carrier immediately upon receipt of the metal.

Another defect which may occur after shipment is *surface stain.* Such stains are caused when a reactive fluid comes in contact with the surface. While oxygen in the air is always reacting with the metal, this oxidation process is slow and the surface gradually darkens over a period of months. However, in the presence of moisture this reaction speeds up and damp spots will darken rapidly. On rare occasions metal may not be properly dried during the final mill cleaning operation. If the metal is slit to width and packed with the film of moisture on the surfaces, it will become stained within days. A more frequent cause for such stains is condensation. During transit in a cold climate the strip is cooled. Subsequently, it may be delivered into a warm, humid plant. Moisture will condense on the cold surface, and without sufficient air circulation to dry it quickly, the metal will be stained. These stains are difficult to remove, and every effort should be made to avoid them.

Other sources of stain sometimes encountered are wet lumber used in packing containers, paper containing free sulfur, contaminated or sulfur-containing lubricants, and extraneous fumes from chemical substances which may be in the area where the metal is stored, or be encountered in transit.

A vivid illustration of how unexpected chemical fumes can damage metal occured in a plant that was producing automobile thermostats from brass. The drawn parts contained residual stresses prior to finishing, and occasionally there would be a siege of stress-corrosion cracking of in-process parts. After diligent detective work, a detergent-producing plant, which used ammonia, was found in the area. When the wind was in the proper direction, moist ammonia fumes were blown into the thermostat plant, causing the brass parts to fail.

To avoid staining copper alloys during manufacture, transit, and storage, care should be taken to protect them from contact with liquids, gases, or fumes of materials that react with them. Copper and copper alloys are among the noblest metals, but the films that form on their surfaces, while usually very protective, are dark in color and difficult to remove.

Another class of surface defects is essentially subsurface in origin. Fortunately these defects are controlled in number and severity in modern mill practices, so they occur very rarely, as unusual nuisances. *Slivers, laminations,* and *blisters* are in this category. They originate as subsurface voids and usually have their beginnings in the casting operation or early stages of rolling. A sliver starts as a small void which becomes elongated during rolling and ends up as a dark and flaky streak on the surface. Certain alloys, such as the phosphor bronzes, tend to contain dissolved gases which cause during solidification the kind of void that develops into a sliver. Laminations are similar to slivers, except that they are larger and originate deeper within the metal. When they become exposed during slitting or blanking, the metal separates into two layers.

Blisters result from tiny voids which originate from a different source. They may be caused by a phase change. When beta phase transforms to alpha phase in brass, the alpha crystals are smaller, so there is a contraction which can leave minute voids. When such voids are near the surface, they may become filled with gas and expand during annealing to form a surface blister.

As noted earlier, all such defects are cause for rejection during mill inspection and are removed prior to further processing or shipment. In rare instances, the user may find an isolated defect, which normally can be discarded in a small piece without significant inconvenience or loss.

TROUBLESHOOTING

Every user of copper and copper alloy sheet and strip will occasionally experience difficulty in fabricating a part. When such a situation develops, it is natural to wonder if some inconsistency in the metal has caused it. Sometimes this is the case, and the cause is recognized. When the cause is not readily identifiable, a thorough investigation of all the elements involved in the part-fabricating process may be called for—the press, the tools, the lubricant, the process, and the metal.

An investigation aimed at solving the problem begins at the press where the part is fabricated. The first step should be to determine, specifically, the details surrounding the difficulty. Often a problem may be described too broadly for analysis. As an example, a brass part might be reported simply to be fracturing during drawing. The investigator needs to know much more than this to define the cause and needs answers to such questions as:

1. When was the problem first encountered?
2. Under what circumstances did the first fracture occur?
3. Does failure occur on every part or intermittently?
4. Does each failed part fail in the same manner?
5. What is the recent history of the tools?
6. What is the history of the lubrication system?

This type of information, combined with appropriate chemical and metallurgical tests, will quickly pinpoint the cause of the problem.

The brass mill technical advisor or marketing engineer regularly assists users in solving fabricating problems. He is available on short notice and is backed by training, experience, and the laboratory facilities needed to assist the user. He can provide valuable assistance in setting up the original metal specifications and in solving any problems that develop after the job goes into production.

As noted earlier, a complete, detailed description of the problem, samples of the metal as received, and samples showing fabricating sequence are needed. With these, the investigator can quickly identify the cause and recommend the appropriate corrective action. There may be more than one potential solution to the problem, and a cooperative approach between the supplier and the user will produce the best and most economical one.

The supplier will do everything in his power to provide metal which embodies the qualities and characteristics most desirable for the user's application. Anything the user does to simplify his requirements will be to his advantage, because it will reduce metal cost and increase the number of potential suppliers. Most mills can produce a wide range of alloys and meet a great variety of special requirements. However, it is to everyone's advantage in the long run if good fabricating methods are practiced so that metal with a minimum of special requirements can be used. A brass mill with a good Marketing Engineering Department can be of valuable help to the user's purchasing staff in establishing effective specifications and weeding out unnecessarily restrictive requirements.

15

ECONOMICS OF
STRIP SELECTION

The three basic elements in the cost of fabricating parts from strip are (1) raw materials; (2) fabricating operations; (3) scrap. The choice of raw material is based on the functional requirements of the part and the length of time it must last. The fabricating method is usually established by the choice of the raw material. The scrap cost is a function of both raw material and fabricating process. This chapter assumes that a copper or copper alloy has already been determined as the material which meets the basic functional criteria for the part to be produced. It goes on to review the factors which influence cost and must be considered when ordering strip or sheet.

Copper alloys often are less costly than any other metal system of comparable properties. In addition, parts fabricated from copper alloy strip, because of excellent formability, are frequently substituted for castings and screw machine parts for economic reasons. The savings to be discussed here, however, do not refer to those just mentioned. Rather, this section will discuss the economies of copper alloy strip resulting from the proper choice of alloy and the efficient use of that strip.

The unit price on a purchase order of copper alloy strip can be expressed in dollars per hundred weight. This cost is typically the sum of several elements, including base price, quantity extra, gauge and width extra, and other special extras. To determine the true cost of metal for a part, other factors must be considered. A prime consideration in the use of strip is the trim scrap usually produced when manufacturing parts. This is resold, and the dollars recouped (not necessarily the same for all alloys) reduce the purchase cost. Where high scrap production is inherent, steel, plated for corrosion resistance, may cost more than a copper alloy because of the low scrap value.

Another factor influencing actual cost is the choice of the thickness and width of strip for the production of a particular part. While the physical requirements must be considered, a judicious choice of gauge and width can frequently effect 5 to 10% cost savings in the initial purchase price. This is so because pricing schedules contain price breaks, based on gauge and width. For example, difference in the gauge and width extras for brass 0.025 × 0.500 and 0.024 × 0.490 may result in cost differnces in this range.

The difference between the above two sizes is, of course, small, but it may be possible at times to justify greater differences in gauge or width, depending upon the scrap value and purchase price relationship. The additional scrap generated may add less cost than the higher price if thinner or narrower strip is ordered. This will have to be evaluated in each individual case. The engineer who establishes strip dimensional requirements needs awareness of pricing practices.

Consistent and uniform quality has a subtle but important effect on the user's cost. As discussed in Chapter 14, because of inherent manufacturing variables, tolerances are required on all properties, including gauge, width, tensile strength, grain size, and even chemical compositions. Products which are supplied with unvarying quality result in lower subsequent manufacturing cost than those which vary within the same tolerance limits in an erratic manner. The supplier who provides the most invariant product may reduce the user's cost more than the supplier with the lowest price.

Simply stated, price alone does not reflect cost or value! Frequently more obscure metal properties, such as density, coil length, directionality, and strength, play a significant role in achieveing value and minimum cost.

EFFECT OF DENSITY ON COST

Occasionally, there is more than one alloy that will suffice for a given application. While these alloys may have similar purchase prices, they can have as much as 8% difference in density. Such differences offer an opportunity for the wise user to decrease his purchasing costs because strip is sold by the pound, while manufactured parts are sold by the piece. An example of the difference in cost caused by a small density difference is given in Table 15–1.

The densities of a number of alloys are given in Table 6–2.

TABLE 15-1
Effect of Density Differences on Net Cost

Assume that the purchase price ($1.50/lb) is the same for Alloys C76200 and
C51000 and that both will perform the job.
Density of C76200—0.310 lb/in.³
Density of C51000—0.320 lb/in.³
Percent difference—Alloy C76200 is 3.1% lower in density than Alloy C51000.
Apparent savings with C76200: 3.1% × $1.50 = 5¢/lb
But scrap value of Alloy C51000 is about 3.5% higher than that of Alloy C76200.
Assume the scrap produced is 50% and the scrap values of Alloys C51000
and C76200 are 77.5 and 75¢/lb, respectively.

Variable	Alloy C76200	Alloy C51000
Amount required to make given number of parts	97	100
Purchase price	$145.50	$150.00
Scrap produced	48.5	50
Scrap value	$ 36.38	$ 38.75
Net cost	$109.12	$111.25
Savings	$ 2.73 = 2.5%	

EFFECT OF COIL LENGTH ON COST

The use of long coils can significantly improve the economy of copper
alloy strip. Brass mills can provide coils heavier than 350 lb./in. of width
in most alloys. Savings result from large coils because fewer are used to
produce the same number of parts. This means that downtime for
changing coils is minimized. Also, larger coils produce measurable
reduction in scrap because of the lesser number of discarded coil ends.
Stopping and starting coils on a frequent basis puts undue strain on
expensive tooling. Experience shows that tool breakage occurs more
frequently during start-ups than at any other time. The use of larger
coils minimizes this danger. Finally, long coils can reduce costs through
lessened manpower requirements by allowing one press man to operate
more than one press.

Even with only a moderate sized facility, if expenditures are required
for larger coil handling equipment they can often be quickly recovered
by the significant savings from processing long coils. This is particularly
true in high production operations.

An accurate prediction of the improved economies with the larger
coils is enlightening. When the alloy, density, gauge, width, coil size,
press speed, progression, downtime to change coils, end scrap per coil,

and total operating cost per machine hour, are known, the desired cost improvement can be calculated with the form shown in Table 15–2.

EFFECT OF DIRECTIONALITY OF PROPERTIES ON COST

A property of all rolled materials which can significantly affect manufacturing cost is the *directionality*. Cold-rolled sheet and strip of all metals exhibit *anisotropy*, that is, a given tensile property such as yield strength or elongation will vary depending on the direction of the specimen with respect to the direction in which the strip was rolled. (See Figure 15–1.)

Such anisotropy results from reduction of the product from a cast slab all the way to sheet or strip by rolling it with the long axis always in the same direction. Working the metal in this fashion tends to align certain atomic or crystallographic planes in each grain so they all point the same way. This orients the slip systems so that deformation in some directions is easier than in others. The resulting mechanical properties, particularly ductility, are different if they are determined with the axis of the speciment parallel to the rolling direction than if measured across the axis of rolling. This means that formability is not the same in all directions, and therefore the orientation of part layout on the strip can be critical.

TABLE 15-2
Cost Savings from Long Coils

I. SPECIFICATIONS

						Regular Coils		Long Coils			
Alloy	Density, lb/in.³	Gauge, in.	Width, in.	I.D., in.	O.D., in.	Weight, lb/in. width	Total Coil Weight, lb	I.D., in.	O.D., in.	Weight, lb/in. width	Total Coil Weight, lb
C26000	0.308	0.051	2.688	16	26	100	265	12	42	394	1050

II. STATISTICAL DATA

    A. Progression (length)   1.7 in.
    B. Press Speed   100 strokes/min
    C. Time Required for Coil Change   15 minutes
    D. End Scrap per Coil
          Tongue   20 in.
          Tail   10 in.
          Total   30 in.
    E. Available Machine Time
          per 8 Hour Shift   360 minutes

TABLE 15-2 *(Cont.)*

III. COMPARISON

A. Coil Length $= \dfrac{\text{Coil Weight}}{\text{Gauge} \times \text{Width} \times \text{Density}} =$ Inches

Regular Coil $= \dfrac{265}{0.051 \times 2.688 \times 0.308} = 6276$ in.

Long Coil $= \dfrac{1050}{0.051 \times 2.688 \times 0.308} = 24{,}868$ in.

B. Lineal Feed = Press Speed $\times$ Progression = Inches per minute
$$= \quad 100 \quad \times \quad 1.7 \quad = 170 \text{ in./min}$$

C. Running Time per Coil $= \dfrac{\text{Coil Length}}{\text{Lineal Feed}} =$ Minutes

Regular Coil $= \dfrac{6276}{170} = 37$ minutes

Long Coil $= \dfrac{24{,}868}{170} = 146$ minutes

D. Total Time per Coil:
Running Time per coil + Time for Change = Minutes
Regular Coil = 37 + 15 = 52 minutes
Long Coil = 146 + 15 = 161 minutes

E. Coils Run per Shift $= \dfrac{\text{Machine Available Time per Shift}}{\text{Total Time per Coil}} =$ Number of Coils

Regular Coil $= \dfrac{380}{52} = 7.3$ Coils per Shift

Long Coil $= \dfrac{380}{161} = 2.3$ Coils per Shift

F. Pieces per Shift $= \dfrac{\text{Coils per Shift} \times (\text{Coil Length} - \text{End Scrap})}{\text{Progression Length}}$
$$= \text{Number of Pieces}$$

Regular Coil $= \dfrac{7.3 \times (6276 - 30)}{1.7} = 26{,}821$ Pieces per Shift

Long Coil $= \dfrac{2.3 \times (24{,}868 - 30)}{1.7} = 33{,}645$ Pieces per Shift

TABLE 15-2 (Cont.)

G. Increase In Production $= \dfrac{\text{Long} - \text{Regular}}{\text{Regular}} \times 100 = \text{Percent}$

$$= \dfrac{33,645 - 26,821}{26,821} \times 100 = 25.4\%$$

H. Machine Cost per 1000 Pieces:

$$\dfrac{\text{Operating Cost}^a \text{per Shift} \times 1000}{\text{Pieces per Shift}} = \text{Dollars}$$

Regular Coil $= \dfrac{\$50.00 \times 1000}{26,821} = \$1.86$

Long Coil $= \dfrac{\$50.00 \times 1000}{33,645} = \$1.49$

I. Machine Downtime per Shift:
Coils per Shift $\times$ Time for Coil Change = Minutes
Regular Coil $= 7.3 \times 15 = 110$ minutes
Long Coil $= 2.3 \times 15 = 35$ minutes

$$\text{Percent Decrease} = \dfrac{(\text{Regular} - \text{Long}) \times 100}{\text{Regular}}$$
$$\dfrac{(110 - 35)}{110} \times 100 = 68.2\%$$

J. Scrap Due to Coil Change (End Scrap):
Density $\times$ Length of End Scrap per Coil $\times$ Width $\times$ Gauge
$0.308 \times 30 \times 2.688 \times 0.051 = 1.27$ lb per coil

K. Metal Saved per Shift Due to Reduced End Scrap:
Pounds per Coil of End Scrap $\times$ Coils Run per Shift = Pounds
Regular Coil $= 1.27 \times 7.3 = 9.3$ lb
Long Coil $= 1.27 \times 2.3 = 2.9$ lb

L. Savings in Scrap $= \dfrac{(\text{Regular} - \text{Long}) \times 100}{\text{Regular}} = \text{Percent}$

$$= \dfrac{(9.3 - 2.9) \times 100}{9.3} = 68.8\%$$

---

^{a}This figure is assumed to cover overhead, wages, rates, etc. Each company must apply its own data to calculate actual savings.

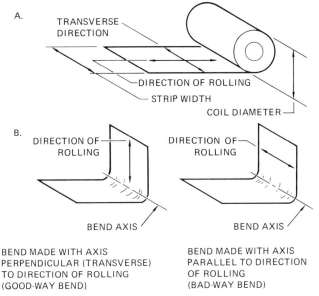

FIGURE 15–1
Nomenclature to describe (A) directionality in strip or sheet and (B) good-way or bad-way bends.

The degree of directionality in properties varies from alloy to alloy, as can be seen from the data tabulated in Table 15–3. This table compares the tensile properties of strip measured in the rolling direction and measured transverse to it.

It is common for parts stamped from strip to be formed by bending to achieve the desired final shape. Bends may be required with axes both transverse and parallel to the rolling direction. Bends made with the axis *transverse to the rolling direction* are denoted as *good-way* bends, and bends made with the bend axis *parallel to the direction of rolling* are denoted as *bad-way* bends. (See Figure 15–1B.) Such terminology historically arose as a consequence of the relatively better good-way bend properties of most metals and single-phase alloys, including brasses, nickel silvers, and phosphor bronzes.

When good-way bends are used to form parts from some hard-rolled alloys, a minimum bend radius of $\frac{1}{32}$ in. may be usable. Should bad-way bends be attempted over this same radius, catastrophic fracture can result. There are some hard-rolled alloys that do not exhibit this extreme directionality, and both good-and bad-way bends can be successfully made with almost the same radius. As a consequence, designers, tool makers, and stampers are afforded some relief from the restrictions

TABLE 15-3
Effect of Testing Direction on the Mechanical Properties of Various Alloys

Alloy	Temper	Orientation[a]	Tension			Elongation, % in 2 in.	Compression	
			$E, \times 10^6$ psi	0.02% Yield Strength, ksi	Ultimate Tensile Strength, ksi		$E, \times 10^6$ psi	0.02% Yield Strength, ksi
26000	Soft	L	16.2	18.0	51.2	57.2	14.5	19.3
	Soft	T	15.8	18.1	50.3	57.8	14.7	19.4
26000	Spring	L	16.0	88.1	96.1	2.7	15.2	52.5
	Spring	T	17.1	83.0	101.5	4.1	17.7	96.1
76200	Soft	L	17.5	34.1	65.7	45.8	18.2	36.1
	Soft	T	18.4	34.6	64.4	47.5	18.6	38.3
76200	Spring	L	18.2	113.3	117.3	1.4	19.2	80.6
	Spring	T	19.2	105.6	120.1	2.3	20.2	122.8
51000	Soft	L	17.2	19.8	48.9	52.8	17.3	21.0
	Soft	T	18.1	19.8	47.9	54.0	17.8	21.2
51000	Spring	L	16.8	103.6	105.4	2.3	16.8	75.3
	Spring	T	18.8	99.6	111.7	4.1	19.7	115.3
19400	Soft	L	17.7	19.7	46.5	30.8	16.4	22.2
	Soft	T	19.0	20.3	45.5	30.7	18.8	23.2
19400	Spring	L	18.4	71.4	73.4	3.7	20.1	55.3
	Spring	T	19.2	67.3	73.1	3.6	20.1	75.0

42500	Soft	L	17.2	22.5	49.1	44.0	16.7	24.9
	Soft	T	17.5	23.6	47.9	47.6	16.4	25.7
42500	Spring	L	17.6	84.1	86.3	4.0	17.1	56.2
	Spring	T	18.8	81.8	93.3	5.5	18.6	91.8
68800	Soft	L	16.4	54.9	84.1	36.4	17.5	57.9
	Soft	T	16.5	55.6	82.1	37.3	17.6	58.7
68800	Spring	L	16.2	111.6	126.9	1.5	15.7	60.8
	Spring	T	16.5	102.9	126.7	2.8	17.2	120.7

[a]L = longitudinal axis of specimen is parallel to the rolling direction; T = longitudinal axis of specimen is transverse to the rolling direction.

related to orientation of part layout with respect to the rolling direction in the strip. This "orientation freedom" is not commonly available, however, because alloys that lack strong directional properties in cold-rolled tempers are few. These relatively nondirectional alloys can be used to decrease the initial cost of strip by allowing better part layout.

An example is the part shown in Figure 15–2 which includes $\frac{1}{32}$ in. radii bends. In order for the part to perform its function, phosphor bronze Alloy C51000 with a nominal tensile strength of 100 ksi was initially considered. The required small bend radii necessitated the layout shown, so the bends would be made the good way (bad-way bends would fracture). Such a layout would have required the purchase of 0.375 in. wide strip material in 0.020 in. gauge. If the part could be reoriented 90°, 2.124 in. wide strip could be used. Manufacturing costs and strip prices vary with the gauge and width (gauge and width extras, as previously mentioned). The particular gauge and width extra for the 0.375 in. wide material originally considered was double the gauge and width extra for the same material at a width of 2.125 in.

Thus a reorientation of 90° as shown in Figure 15–4B, would realize significant economy. Unfortunately, the direction properties of the alloy

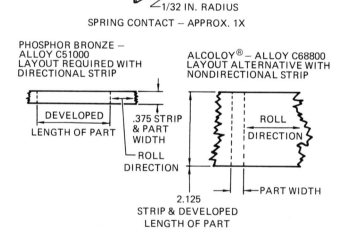

SPRING CONTACT – APPROX. 1X

PHOSPHOR BRONZE –
ALLOY C51000
LAYOUT REQUIRED WITH
DIRECTIONAL STRIP

ALCOLOY® – ALLOY C68800
LAYOUT ALTERNATIVE WITH
NONDIRECTIONAL STRIP

DEVELOPED
LENGTH OF PART

.375 STRIP
& PART
WIDTH

ROLL
DIRECTION

ROLL
DIRECTION

PART WIDTH

2.125
STRIP & DEVELOPED
LENGTH OF PART

1. INITIAL STRIP COST IS LESS FOR 2.125 WIDTH.

2. USE OF 2.125 STRIP REQUIRES FEWER COILS; FEWER COILS PROVIDE ADDITIONAL COST SAVINGS.

FIGURE 15–2
Example of the use of a nondirectional alloy strip to reduce cost of finished parts.

were such that this reorientation was not possible. However, the non-directional properties of another alloy of equivalent strength. Hard temper Alloy C68800, combined with other attributes, allowed the part to be reoriented as desired. Significant cost savings were realized.

Savings in addition to the gauge and width extra reduction in this example were also realized, since less time was lost in stopping and restarting fewer coils. Also more parts could be stamped from the same length of coil, decreasing the number of coils used. This resulted because only a 0.375 in. length of the strip was required per part with the new layout, whereas the old layout required a 2.125 in. length of strip per part.

The freedom from significant directionality in some alloys also provides additional cost savings by web scrap reduction. *Web scrap* is the skeleton of metal generated when parts are blanked from strip. Anything that can be done to minimize this scrap will decrease material cost. Alloys that have little directionality allow parts to be *nested,* or fitted within each other, on the strip to minimize the generation of web scrap. The so-called higher *metal efficiency* allows fewer pounds to be purchased to produce the same number of parts.

An example is the part shown in Figure 15–3. This part includes both good-way and bad-way bends when blanked as shown in Figure 15–4B. If directionality was present because of the temper required, for example, phosphor bronze Alloy C51000 in Spring temper, the layout shown in Figure 15–4A might be required. The 45° bend orientation is often used with highly directional strip because there is less tendency to fracture

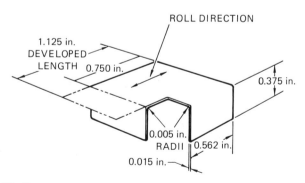

FIGURE 15–3
This hypothetical part has equal radius bends in directions both with and across the rolling direction. Selection of the proper copper strip alloy for this application will depend not only on the material's strength but also on its bend properties in these directions.

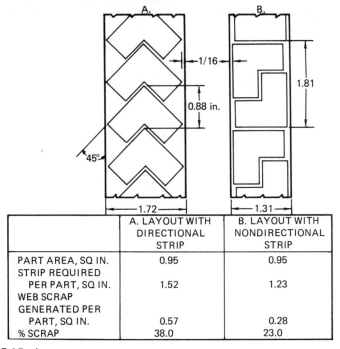

	A. LAYOUT WITH DIRECTIONAL STRIP	B. LAYOUT WITH NONDIRECTIONAL STRIP
PART AREA, SQ IN.	0.95	0.95
STRIP REQUIRED PER PART, SQ IN.	1.52	1.23
WEB SCRAP GENERATED PER PART, SQ IN.	0.57	0.28
% SCRAP	38.0	23.0

FIGURE 15–4

Layout *A* is required for forming directional copper strip alloys, such as Alloy C51000. Selecting a nondirectional alloy, such as C68800, would allow the more efficient layout *B*, with a reduction in web scrap from 38 to 23%.

than with a 90° bend orientation. The most desirable layout, Figure 15-4*B*, would not be possible, because rupture would occur on bad-way bends because of directionality.

The layout required with directional strip generates 38% scrap. Only 62 lbs. of parts are produced for each 100 lbs. of strip purchased. The remaining web scrap represents an immediate loss or degrading of value. This *degradation,* the difference between purchase price and scrap value, will frequently exceed 50% of the purchase price. A significant economic penalty can result from poor metal efficiency. Orienting parts to minimize web scrap is not always possible, regardless of the alloy selected. The plane geometry involved in the layout is often quite simple, but designing tools to do the job is not.

With alloys such as C68800 that exhibit minimum directionality, there is less restriction. With this essentially nondirectional alloy, the part layout could be reoriented as shown in Figure 15–4*B*. Subsequent web scrap produced would be only 23% for a reduction of 40% in waste.

EFFECT OF ALLOY STRENGTH ON COST

Strength is basically a measure of a metal's resistance to deformation. It is one of the desirable characteristics of all metal parts, and there would be no limitation placed on strength were it not that high strength metals are difficult to form. Strength is a primary criterion for selecting a particular metal or alloy, but other properties also must be considered. Combinations of properties are usually needed, such as strength, formability, and corrosion resistance—or, strength, solderability, and corrosion resistance. Finally, cost is always a criterion along with the mechanical and physical requirements.

Although copper is a relatively high cost metal compared to some other common elements such as iron, it combines such useful physical and mechanical properties that its true cost per unit of returned value is low. For example, to use an iron alloy such as a stainless steel, with corrosion resistance equivalent to that of copper, as an electrical conductor would require a cross section so many times greater than the copper one that the cost would be very high by comparison.

The strength of pure copper is readily increased by solid-solution hardening. This is the most common means of producing copper alloys of high strength. The exceptions are the copper-beryllium alloys, which are precipitation strengthened by heat treatment. There are other alloys containing elements whose solubility decreases with temperature, but none is strengthened as much as copper-beryllium by precipitation heat treatment. Copper-beryllium alloys are very costly compared to solid-solution strengthened alloys such as the phosphor bronzes, nickel silvers, and high zinc brasses, but because of their very high strength and hardness they have no rivals for some special applications.

High zinc brass Alloy C26000, cartridge brass, has good strength, good formability, good resistance to atmospheric corrosion, and good conductivity, and its cost is the lowest of the standard copper alloys. All of these valuable characterists make Alloy C26000 the logical least cost choice for many parts whose functions require some strength. Indeed Alloy C26000 is by far the most popular copper alloy because of this combination of good strength and low cost. One disadvantageous characteristic of this alloy is that it is subject to stress corrosion by atmospheres containing ammonia. Residual stresses left in parts from forming operations in their manufacture can be reduced below stress-corrosion threshold levels by stress-relieving, low temperature heat treatments. Alternatively, brass can be plated to protect it from such atmospheres. When neither of these solutions is feasible, another alloy must be chosen.

All solid-solution strengthened copper alloys other than the brasses contain addition elements of higher cost than zinc, or are more difficult to process than the brasses, and therefore cost more. Two alloys which have been used to replace C26000 where greater resistance to corrosion and stress corrosion is needed without significantly sacrificing strength are C42200 (Cu—87%, Zn—12%, Sn—1%), Olin Lubronze,[1] and C42500 (Cu—88.5%, Zn—9.5%, Sn—2%), Olin Lubaloy X.[1] Their strength and formability are similar to those of cartridge brass, and they are roughly 10 and 20% higher in price,[2] respectively, than Alloy C26000.

The nickel silvers are essentially brasses, that is, copper-zinc alloys, but nickel has been used to replace some of the copper. They have a silver color, and are stronger than the brasses with the same zinc content. The nickel provides a mild additional solid-solution strengthening effect to that of the zinc. Tarnish resistance, oxidation resistance, corrosion resistance, and color are factors in the choice of nickel silvers. Information on these characteristics is contained in Chapters 8 and 12. Our concern here is their relative strength and cost. Alloy C76200 (Cu—59%, Zn—29%, Ni—12%) is 15 to 20% stronger than C26000 in each rolled temper, and C77000 (Cu—55%, Zn—27%, Ni—18%) is a little stronger than C762000. Both of these alloys have been used extensively in relay springs, which are flat or moderately bent, because of their high strength and stiffness. Their modulus of elasticity, a measure of stiffness, or resistance to deflection is 18×10^6 psi as compared to 16×10^6 psi, or a little less, for most copper alloys. In base price they are 30 to 40% higher than Alloy C26000, and their gauge and width extras are also greater, so their cost is considerably higher. Yet, where surfaces must be resistant to tarnishing and oxidation, as in contact spring applications requiring high strength, their value has often been worth the extra cost.

There are three popular copper-tin alloys, phosphor bronzes, which are used for their combined strength, formability, corrosion resistance, and fatigue strength. Alloy C51100 (Cu—96%, Sn—4%, P) has strength and formability comparable to those of Alloy C26000 and is very resistant to stress corrosion and corrosion fatigue. It is used frequently in electrical terminals and connectors where its 20% IACS electrical conductivity is needed along with high functional reliability. Alloy C51000 (Cu—95%, Sn—5%, P) has higher strength and better fatigue life, corrosion resistance, and formability, but lower electrical conductivity. Both alloys are priced close to 50% higher than Alloy C26000. Alloy C52100 (Cu—92%, Sn—8%, P) is a very high strength phosphor bronze.

---

[1] Trademark of Olin Corporation,
[2] Cost and price comparisons are estimates based on metal values at this time, and they can change with changes in metal values, so they are approximations.

Its strength is approximately 20% higher than that of C26000, and it has high fatigue strength and corrosion resistance. It is frequently used in spring applications where the part must not fail under repeated cyclical stress. Alloy C51000 is also used in such applications. Alloy C52100 costs approximately 60% more than C26000.

Because strength is often most important in working parts fabricated from copper and copper alloys, metallurgical means other than solid-solution strengthening have been used to develop alloys that fill many of the property gaps that existed when only the solid-solution strengthened alloys were available. When soluble elements are used to strengthen copper, electrical conductivity suffers. Heat-treatable precipitation-strengthening alloys tend to be high priced and difficult to use. A good compromise was the development of the copper-iron-phosphorus alloys and a process that produced high conductivity without special heat treatments. The most popular strip alloy, C19400 (Cu—97%, Fe—2%, P. Zn), combines strengthening by a fine dispersion of intermetallic particles with some solid-solution hardening. The strength of C19400 is 50% greater than that of copper, its electrical conductivity is only 25% less, and its price is lower. Further, because of its combined strength and conductivity, it can be applied to reduce the cross sections of sectional thickness to allow parts to carry greater electric current without overheating to increase their usefulness. A comparison of Alloy C19400 with some precipitation-strengthened alloys is given in Table 15–4.

The table includes those alloys which have been marketed as strip products. Their strengths, cost indices, and conductivities are pertinent.

Another alloy which is based on the copper-iron-phosphorus system is Alloy C19500 (cu—97%, Fe—1.5%, Co—0.8%, Sn—0.5%, P). This alloy has the highest strength of the modified coppers that do not require precipitation heat treatments. Its strength is almost double and its conductivity is 50% of that of copper, while its cost is a little more than 10% higher.

Another group of alloys which depend on the combination of solid-solution strengthening and dispersion strengthening has developed as high strength, low cost alloys. These are for applications, in which electrical conductivity is not a critical requirement, but strength, forma-bility, and cost are. Earlier in this chapter the cost savings potential of alloys which exhibited a minimum of directionality in rolled tempers was pointed out. The alloys referred to were those in which a dispersion-strengthening mechanism was in part responsible for the alloy's strength when cold rolled. Dispersion strengthening is nondirectional.

Alloy C66400 (Cu—86.5%, Zn—11.5%, Fe—1.5%, Co—0.5%) is about 10% stronger than Alloys C26000, C42200 and C42500. It has excellent formability and good bad-way bend properties coupled with

TABLE 15-4
Comparison of Conductivity, and Strength of Olin Alloy C19400 with Other Special high Strength, High Conductivity, Premium Priced Alloys

Alloy No.	C19400	C15500	C16200	C17500	C64700
Name	HSM Copper	Cu-Mg-P "SS-Cu"	Cadmium Copper	Alloy 10 Beryllium Copper	(Various) Copper-Nickel Silicon Alloy
Conductivity (% IACS)	65-75%	85%	87%	20-60% (a)	22-42% (a)
Rolled Tempers Heat Treatment Required	None	None	Yes	Yes (a)	Yes (a) (b)
Yield Strength (0.2% Offset), ksi					
1/2 Hard	46	41	46	60(110)	48(89)
Hard	61	57	57	70(125)	57(94)
Extra Hard	68	65	62	–	63(96)
Spring	71	71	67	–	65(97)
Tensile Strength, ksi					
1/2 Hard	58	53	48	67(120)	74(102)
Hard	65	62	57	70(110)	83(104)
Extra Hard	70	67	65	–	88(104)
Spring	73	72	69	–	90(104)
Annealed Tempers Yield Strength (0.2% Offset), ksi					
Special Light Anneal	47 ⎫				
Light Anneal	23 ⎬	18	12	AT 90	16(73)
Soft	22 ⎭				
Tensile Strength, ksi					
Special Light Anneal	58 ⎫				
Light Anneal	50 ⎬	40	37	AT 110	40(89)
Soft	45 ⎭				

(a) Data shown is for cold rolled and heat treated conditions. Properties of these alloys without heat treatment are considerably poorer. Strength and conductivity are lower. Heat treated properties are in parentheses.
(b) Data quoted is for rod, strip properties lower.

high strength. It can replace C42200 at equal cost, and provide higher strength, better formability, and equal corrosion resistance and conductivity. It can replace Alloy C42500 at 5% lower cost, higher strength, better formability, and equal corrosion resistance and conductivity.

Alloy C68800 (Cu—73%, Zn—23%, Al—3.5%, Co—.4%) is a very high strength, low cost brass. Its base price is more than 20% less and its gauge and width extras are less than those for Alloy C76200. Its strength is 10% or more above that of Alloy C76200, and its formability is equivalent. It can be substituted for C76200 or C77000 at considerably lower cost. Its base price is approximately 60% of that of Alloy C51000, and its gauge and width extras are also lower. The alloy has greater strength than 5% tin phosphor bronze and good formability, making it a viable substitute in many applications with a good potential cost savings.

One more high strength alloy should be mentioned, because it too offers a strength advantage at lower cost, when compared to the phosphor bronzes and high strength nickel silvers. Alloy C63800 (Cu—95%, Al—2.8%, Sil—1.8%, Co—0.4%) in rolled tempers is from 10 to 20% stronger than the standard high strength phosphor bronzes and nickel silvers, and its base price is 30 to 40% lower, making it a potentially cost saving substitute for these other alloys in many applications.

All the preceding information has centered primarily on strength, fatigue resistance, electrical conductivity, and cost. There are still other properties that must be weighed. Stress relaxation characteristics, as pointed out in Chapters 7 and 13, show that material selection cannot be made on strength determination alone. There is no one material that is completely satisfactory for all situations. An engineer or designer must determine where the trade-offs can be made.

METHODS FOR CALCULATING COMPARATIVE COSTS

In order to obtain the maximum economic advantage of copper alloy strips, it is frequently necessary to choose between two or more competing alloys, not necessarily of the same dimensions. If all the mechanical, physical, and chemical requirements of the application are satisfied by

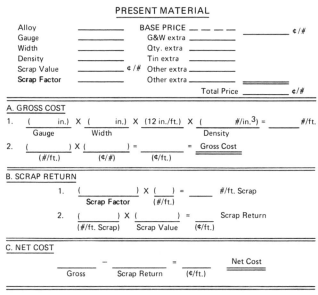

FIGURE 15–5
Form for material cost comparison.

PROPOSED MATERIAL

Alloy _____   BASE PRICE _ _ _ _ _____ ¢/#
Gauge _____   G&W extra _____
Width _____   Qty. extra _____
Density _____   Tin extra _____
Scrap Value _____ ¢/# Other extra _____
Scrap Factor _____   Other extra _____
                                    Total Price _____ ¢/#

A. GROSS COST

1. (_____ in.) X (_____ in.) X (12 in./ft.) X (_____ #/in.^{3}) = _____ #/ft.
         Gauge   Width   Density

2. (_____) X (_____) = _____ = Gross Cost
      (#/ft.)   (¢/#)   (¢/ft.)

B. SCRAP RETURN

1. (_____) X (_____) = _____ #/ft. Scrap
      Scrap Factor   (#/ft.)

2. (_____) X (_____) = _____ Scrap Return
      (#/ft. Scrap) Scrap Value   (¢/ft.)

C. NET COST

_____ — _____ = _____ Net Cost
  Gross   Scrap Return   (¢/ft.)

D. PERCENT SAVINGS

(Net Cost Present) — (Net Cost Proposed) X 100 = _____ %
_____   Savings
        (Net Cost Present)

FIGURE 15-5
(Continued)

these materials, then the choice becomes an economic one. Since there is a difference in initial selling price, density, and scrap value, a *net cost* must be calculated.

The net cost per foot of strip for each alloy considered can be calculated by the simple equation

where

$$N = 12\,GW\,\theta(P\text{-}jS)$$
$$N = \text{net cost per foot } (\cent)$$
$$G = \text{gauge (in.)}$$
$$W = \text{width (in.)}$$
$$\theta = \text{density (lb/in.}^3)$$
$$P = \text{selling price } (\cent/\text{lb or \$/cwt})$$
$$j = \text{scrap generated } (\%/100)$$
$$S = \text{scrap price } (\cent/\text{lb or \$/cwt})$$

A material cost comparison form (Figure 15–5) can be used in lieu of the "cost equation." It is especially useful in showing the relative cost elements. Presentation of cost data is also facilitated via the form.

APPENDIX

WROUGHT COPPER AND COPPER ALLOYS LIST[a]

COPPERS

Composition, % max. unless otherwise indicated

Copper Number	Desig-nation	Description	Cu (incl. Ag), % min.	Ag % min.	Ag Troy oz min.	As	Sb	P	Te	Other Named Elements
C10100[1]	OFE	Oxygen-Free Electronic	99.99[2]	—	—	(3)	(3)	.0003	.0010	(3)
C10200[1]	OF	Oxygen-Free	99.95	—	—	—	—	—	—	—
C10300	OFXLP		99.95[4]	—	—	—	—	.001–.005	—	—
C10400[1]	OFS	Oxygen-Free with Ag	99.95	.027	8	—	—	—	—	—
C10500[1]	OFS	Oxygen-Free with Ag	99.95	.034	10	—	—	—	—	—
C10700	OFS	Oxygen-Free with Ag	99.95	.085	25	—	—	—	—	—
C10800	OFLP		99.95[4]	—	—	—	—	.005–.012	—	—
C10920	—		99.90	.044	13	—	—	—	—	.02 Oxygen
C10930	—		99.90	.085	25	—	—	—	—	.02 Oxygen
C10940	—		99.90	—	—	—	—	—	—	.02 Oxygen
C11000[1]	ETP	Electrolytic Tough Pitch	99.90	—	—	—	—	—	—	—
	FRHC	Fire Refined High Conductivity	99.90	—	—	—	—	—	—	—
	CRTP	Chemically Refined Tough Pitch	99.90	—	—	—	—	—	—	—
C11100[1]	—	Electrolytic Tough Pitch Anneal Resistant	99.90	—	—	—	—	—	—	(5)
C11300[1][6]	STP	Tough Pitch with Ag	99.90	.027	8	—	—	—	—	—
C11400[1][6]	STP	Tough Pitch with Ag	99.90	.034	10	—	—	—	—	—
C11500[1][6]	STP	Tough Pitch with Ag	99.90	.054	16	—	—	—	—	—
C11600[1][6]	STP	Tough Pitch with Ag	99.90	.085	25	—	—	—	—	—
C11700	—		99.90[7]	—	—	—	—	.04	—	.004–.02B
C12000	DLP	Phosphorus Deoxidized, Low Residual Phosphorus	99.90	—	—	—	—	.004–.012	—	—
C12100	—		99.90	.014	4	—	—	.005–.012	—	—
C12200[8]	DHP	Phosphorus Deoxidized, High Residual Phosphorus	99.9	—	—	—	—	.015–.040	—	—
C12300	—		99.90	.014	4	—	—	.015–.040	—	—
C12500[9]	FRTP	Fire Refined Tough Pitch	99.88	—	—	.012	.003	—	.025[10]	.050Ni .003Bi .004Pb

Alloy No.	Designation	Name	Cu min %							Other elements
C12700[9]	FRSTP	Fire Refined Tough Pitch with Ag	99.88	.027	8	.012	.003	—	.025[10]	.050Ni .003Bi .004Pb
C12800[9]	FRSTP	Fire Refined Tough Pitch with Ag	99.88	.034	10	.012	.003	—	.025[10]	.050Ni .003Bi .004Pb
C12900[9]	FRSTP	Fire Refined Tough Pitch with Ag	99.88	.054	16	.012	.003	—	.025[10]	.050Ni .003Bi .004Pb
C13000[9]	FRSTP	Fire Refined Tough Pitch with Ag	99.88	.085	25	.012	.003	—	.025[10]	.050Ni .003Bi .004Pb
C14200	DPA	Phosphorus Deoxidized, Arsenical	99.4	—	—	.15–.50	—	.015–.040	—	—
C14300	—	Cadmium Copper, Deoxidized	99.90[11]	—	—	—	—	—	—	.05–.15Cd
C14310	—		99.90[11]	—	—	—	—	—	—	.10–.30Cd
C14500[12]	DPTE	Phosphorus Deoxidized, Tellurium-Bearing	99.90[13]	—	—	—	—	.004–.012[14]	.40–.6	—
C14700	—	Sulfur-Bearing	99.90[13]	—	—	—	—	—	—	.20–.50S
C14710	—		99.90[15][16]	—	—	—	—	.010–.030	—	.20–.50S .10Pb
C14720	—		99.50[15][16]	—	—	—	—	.010–.030	—	.05–.15S .05Pb
C15000	—	Zirconium Copper	99.80	—	—	—	—	—	—	.10–.20Zr
C15500	—		99.75	.027–.10	8–30	—	—	.040–.080	—	.08–.13Mg
C15710	—		99.80[17]	—	—	—	—	—	—	.15–.25Al₂O₃ .01Fe .01Pb
C15720	—		99.80[17]	—	—	—	—	—	—	.04 Oxygen .35–.45Al₂O₃ .01Fe .01Pb
C15735	—		99.80[17]	—	—	—	—	—	—	.04 Oxygen .65–.75Al₂O₃ .01Fe .01Pb .04 Oxygen

HIGH COPPER ALLOYS

Copper Alloy Number	Previous Trade Name	Composition, % max. unless shown as a range or minimum										
		(18) Cu (incl. Ag) + Elements with Specific Limits, % min.	Fe	Sn	Ni	Co	Cr	Si	Be	Pb	Cd	Other Named Elements
C16200	Cadmium Copper	99.8	.02	.50-.7	—	—	—	—	—	—	.7-1.2	—
C16500		99.8	.02	—	—	—	—	—	—	—	.6-1.0	—
C17000	Beryllium Copper	99.5[2]	(19)	—	(19)	(19)	—	—	1.60-1.79	—	—	—
C17200	Beryllium Copper	99.5[2]	(19)	—	(19)	(19)	—	—	1.80-2.00	—	—	—
C17300		99.5[2]	(19)	—	(19)	(19)	—	—	1.80-2.00	.20-.6	—	—
C17500	Beryllium Copper	99.5[2]	.10	—	—	2.4-2.7	—	—	.40-.7	—	—	.9-1.1Ag
C17600		99.5[2]	.10	—	—	1.4-1.7	—	—	.25-.50	—	—	.40-.6Te
C17700		99.5[2]	.10	—	—	2.4-2.7	—	—	.40-.7	—	—	—
C18200	Chromium Copper	99.5	.10	—	—	—	.6-1.2	.10	—	.05	—	.005As
C18400	Chromium Copper	99.8	.15	—	—	—	.40-1.2	.10	—	—	—	.005Ca .05Li .05P .7Zn
C18500	Chromium Copper	99.8	—	—	—	—	.40-1.0	—	—	.015	—	.04P
C18700		99.9	—	.6-.9	—	—	—	—	—	.8-1.5	—	.08-.12Ag
C18900		99.9	—	.6-.9	—	—	—	.15-.40	—	.02	—	.05P .01Al .10-.30Mn .10Zn
C19000		99.5	.10	—	.9-1.3	—	—	—	—	.05	—	.8Zn .15-.35P
C19100		99.5	.20	—	.9-1.3	—	—	—	—	.10	—	.50Zn .35-.6Te .15-.35P

Composition, % max. unless shown as a range or minimum

Copper Alloy Number	Cu, % min.	Fe	Sn	Zn	Al	Pb	P	Co
C19200	98.7[20]	.8–1.2	—	—	—	—	.01–.04	—
C19400	97.0[20]	2.1–2.6	—	.05–.20	—	.03	.015–.15	—
C19500	96.0[20]	1.3–1.8	.40–.7	.20	.02	.02	.08–.12	.6–1.1
C19600	99.7 min	.9–1.2	—	.35	—	—	.25–.35	—

COPPER–ZINC ALLOYS (BRASSES)

Copper Alloy Number	Previous Trade Name	Composition, % max. unless shown as a range					Other Named Elements
		Cu	Pb	Fe	Zn[20]	P/wf/	
C20500		97.0–98.0	.02	.05	Rem.	—	—
C21000	Gilding, 95%	94.0–96.0	.05	.05	Rem.	—	—
C22000	Commercial Bronze, 90%	89.0–91.0	.05	.05	Rem.	—	—
C22600	Jewelry Bronze, 87½%	86.0–89.0	.05	.05	Rem.	—	—
C23000	Red Brass, 85%	84.0–86.0	.05	.05	Rem.	—	—
C23030		83.5–85.5	.05	.05	Rem.	—	.20–.40Si
C23400		81.0–84.0	.05	.05	Rem.	—	—
C24000	Low Brass, 80%	78.5–81.5	.05	.05	Rem.	—	—
C25000		74.0–76.0	.05	.05	Rem.	—	—
C26000	Cartridge Brass, 70%	68.5–71.5	.07	.05	Rem.	—	—
C26100		68.5–71.5	.05	.05	Rem.	.02–.05	—
C26200		67.0–70.0	.07	.05	Rem.	—	—
C26800	Yellow Brass, 66%	64.0–68.5	.15	.05	Rem.	—	—
C27000	Yellow Brass, 65%	63.0–68.5	.10	.07	Rem.	—	—
C27200		62.0–65.0	.07	.07	Rem.	—	—
C27400	Yellow Brass, 63%	61.0–64.0	.10	.05	Rem.	—	—
C28000	Muntz Metal, 60%	59.0–63.0	.30	.07	Rem.	—	—
C28200		58.0–61.0	.03	.05	Rem.	.12–.22	.005Al[21] .05Sn

COPPER–ZINC–LEAD ALLOYS (LEADED BRASSES)

Copper Alloy Number	Previous Trade Name	Composition, % max. unless shown as a range					Other Named Elements
		Cu	Pb	Fe	Sn	Zn[20]	
C31400	Leaded Commercial Bronze	87.5–90.5	1.3–2.5	.10	—	Rem.	.7Ni
C31600	Leaded Commercial Bronze (Nickel-Bearing)	87.5–90.5	1.3–2.5	.10	—	Rem.	.7–1.2Ni .04–.10P
C32000	Leaded Red Brass	83.5–86.5	1.5–2.2	.10	—	Rem.	.25Ni
C33000	Low Leaded Brass (Tube)	65.0–68.0	.20–.8[22]	.07	—	Rem.	—
C33100		65.0–68.0	.7–1.2	.06	—	Rem.	—
C33200	High Leaded Brass (Tube)	65.0–68.0	1.3–2.0	.07	—	Rem.	—
C33500	Low Leaded Brass	62.5–66.5	.30–.8	.10	—	Rem.	—
C34000	Medium Leaded Brass, 64½%	62.5–66.5	.8–1.4	.10	—	Rem.	—
C34200	High Leaded Brass, 64½%	62.5–66.5	1.5–2.5	.10	—	Rem.	—
C34400		62.0–66.0	.50–1.0	.10	—	Rem.	—
C34500		62.0–64.0	1.5–2.8	.10	—	Rem.	—
C34700		62.5–64.5	1.0–1.8	.10	—	Rem.	—
C34800		61.5–63.5	.40–.8	.10	—	Rem.	—
C34900		61.0–64.0	.10–.50	.10	—	Rem.	—
C35000	Medium Leaded Brass, 62%	59.0–61.0[23]	.8–1.4	.10	—	Rem.	—
C35300	High Leaded Brass, 62%	59.0–64.5[23]	1.3–2.3	.10	—	Rem.	—
C35600	Extra High Leaded Brass	59.0–64.5[24]	2.0–3.0	.10	—	Rem.	—
C36000	Free Cutting Brass	60.0–63.0	2.5–3.7	.35	—	Rem.	—
C36200		60.0–63.0	3.5–4.5	.15	—	Rem.	—
C36500	Leaded Muntz Metal, Uninhibited	58.0–61.0	.40–.9	.15	.25	Rem.	—
C36600	Leaded Muntz Metal, Arsenical	58.0–61.0	.40–.9	.15	.25	Rem.	.02–.10As
C36700	Leaded Muntz Metal, Antimonial	58.0–61.0	.40–.9	.15	.25	Rem.	.02–.10Sb
C36800	Leaded Muntz Metal, Phosphorized	58.0–61.0	.40–.9	.15	.25	Rem.	.02–.10P

COPPER–ZINC–LEAD ALLOYS (LEADED BRASSES) (Cont.)

Copper Alloy Number	Previous Trade Name	Composition, % max. unless shown as a range						
		Cu	Pb	Fe	Sn	Zn[20]		Other Named Elements
C37000	Free Cutting Muntz Metal	59.0–62.0	.9–1.4	.15	—	Rem.		—
C37100		58.0–62.0	.6–1.2	.15	—	Rem.		—
C37700	Forging Brass	58.0–61.0	1.5–2.5	.30	—	Rem.		—
C37800		57.0–60.0	1.0–2.5	.30	—	Rem.		—
C38000		55.0–60.0	1.5–2.5	.35	.30	Rem.		.50Al
C38500	Architectural Bronze	55.0–60.0	2.0–3.8	.35	—	Rem.		—
C38590		56.5–60.0	2.0–3.5	.35	—	Rem.		—
C38600		56.0–59.0	2.5–4.5	.35	—	Rem.		.02Sb

COPPER–ZINC–TIN ALLOYS (TIN BRASSES)

Copper Alloy Number	Previous Trade Name	Composition, % max. unless shown as a range								Other Named Elements
		Cu	Pb	Fe	Sn	Zn[20]	P	As	Sb	
C40500		94.0–96.0	.05	.05	.7–1.3	Rem.	—	—	—	—
C40800		94.0–96.0	.05	.05	1.8–2.2	Rem.	—	—	—	—
C41000		91.0–93.0	.05	.05	2.0–2.8	Rem.	—	—	—	—
C41100		89.0–92.0	.10	.05	.30–.7	Rem.	—	—	—	—
C41300		89.0–93.0	.10	.05	.7–1.3	Rem.	—	—	—	—
C41500		89.0–93.0	.10	.05	1.5–2.2	Rem.	—	—	—	—
C42000		88.0–91.0	—	—	1.5–2.0	Rem.	.25	—	—	—
C42100		87.5–89.0	.05	.05	2.2–3.0	Rem.	.35	—	—	.15–.35Mn
C42200		86.0–89.0	.05	.05	.8–1.4	Rem.	.35	—	—	—
C42500		87.0–90.0	.05	.05	1.5–3.0	Rem.	.35	—	—	—
C43000		84.0–87.0	.10	.05	1.7–2.7	Rem.	—	—	—	—
C43200		85.0–88.0	.05	.05	.40–.6	Rem.	.35	—	—	—
C43400		84.0–86.0	.05	.05	.50–1.0	Rem.	—	—	—	—
C43500		79.0–83.0	.10	.05	.6–1.2	Rem.	—	—	—	—
C43600		80.0–83.0	.05	.05	.20–.50	Rem.	—	—	—	—
C44300	Admiralty, Arsenical	70.0–73.0	.07	.06	.9–1.2[25]	Rem.	—	.02–.10	—	—
C44400	Admiralty, Antimomial	70.0–73.0	.07	.06	.9–1.2[25]	Rem.	—	—	.02–.10	—
C44500	Admiralty, Phosphorized	70.0–73.0	.07	.06	.9–1.2[25]	Rem.	.02–.10	—	—	—
C46200	Naval Brass, 63½%	62.0–65.0	.20	.10	.50–1.0	Rem.	—	—	—	—
C46400	Naval Brass, Uninhibited	59.0–62.0	.20	.10	.50–1.0	Rem.	—	—	—	—
C46500	Naval Brass, Arsenical	59.0–62.0	.20	.10	.50–1.0	Rem.	—	.02–.10	—	—
C46600	Naval Brass, Antimonial	59.0–62.0	.20	.10	.50–1.0	Rem.	—	—	.02–.10	—
C46700	Naval Brass, Phosphorized	59.0–62.0	.20	.10	.50–1.0	Rem.	.02–.10	—	—	—
C47000	Naval Brass Welding and Brazing Rod	57.0–61.0	.05	—	.25–1.0	Rem.	—	—	—	.01Al
C47600		86.0–88.0	1.8–2.2	.05	1.8–2.2	Rem.	.03–.07	—	—	.05–.15Mn
C48200	Naval Brass, Medium Leaded	59.0–62.0	.40–1.0	.10	.50–1.0	Rem.	—	—	—	—
C48500	Naval Brass, High Leaded	59.0–62.0	1.3–2.2	.10	.50–1.0	Rem.	—	—	—	—

COPPER–TIN ALLOYS (PHOSPHOR BRONZES)

Copper Alloy Number	Previous Trade Name	Composition, % max. unless shown as a range						
		Cu + Sn + P, % min.	Pb	Fe	Sn	Zn	P	Al
C50100		99.5	.05	.05	.50–.8	—	.01–.05	—
C50200		99.5	.05	.10	1.0–1.5	—	.04	—
C50500	Phosphor Bronze, 1.25% E	99.5	.05	.10	1.0–1.7	.30	.03–.35	—
C50700		99.5	.05	.10	1.5–2.0	—	.30	—
C50800		99.5	.05	.10	2.6–3.4	—	.01–.07	—
C50900		99.5	.05	.10	2.5–3.8	.30	.03–.30	—
C51000	Phosphor Bronze, 5% A	99.5	.05	.10	4.2–5.8	.30	.03–.35	—
C51100		99.5	.05	.10	3.5–4.9	.30	.03–.35	—
C51800	Phosphor Bronze	99.5	.02	—	4.0–6.0	—	.10–.35	.01
C51900		99.5	.05	.10	5.0–7.0	.30	.03–.35	—
C52100	Phosphor Bronze, 8% C	99.5	.05	.10	7.0–9.0	.20	.03–.35	—
C52400	Phosphor Bronze, 10% D	99.5	.05	.10	9.0–11.0	.20	.03–.35	—

COPPER–TIN–LEAD ALLOYS (LEADED PHOSPHOR BRONZES)

Copper Alloy Number	Previous Trade Name	Composition, % max. unless shown as a range					
		Cu + Sn + P + Pb, % min.	Pb	Fe	Sn	Zn	P
C53200	Phosphor Bronze B	99.5	2.5–4.0	.10	4.0–5.5	.20	.03–.35
C53400	Phosphor Bronze B-1	99.5	.8–1.2	.10	3.5–5.8	.30	.03–.35
C54400	Phosphor Bronze B-2	99.5[26]	3.5–4.5	.10	3.5–4.5	1.5–4.5	.01–.50
C54800		99.5[26]	4.0–6.0	.10	4.0–6.0	.30	.03–.35

COPPER–ALUMINUM ALLOYS (ALUMINUM BRONZES)

Composition, % max. unless shown as a range or minimum

Copper Alloy Number	(18) Cu + Elements with Specific Limits, % min.	Cu (incl. Ag)	Pb	Fe	Sn	Zn	Al	As	Mn	Si	Ni (incl. Co)	Co	P
C60600	99.5	92.0–96.0	—	.50	—	—	4.0–7.0	—	—	—	—	—	—
C60700	99.5	94.6–96.0	.01	—	1.7–2.0	—	2.3–2.9	—	—	—	—	—	—
C60800	99.5	92.5–94.8	.10	.10	—	—	5.0–6.5	.02–.35	—	—	—	—	—
C61000	99.5	90.0–93.0	.02	.50	—	.20	6.0–8.5	—	—	.10	—	—	—
C61300	99.5	86.5–93.8	—	3.5	.20–.50	—	6.0–8.0	—	—	—	—	—	—
C61400	99.5	88.0–92.5	.01	1.5–3.5	—	.20	6.0–8.0	—	.50	—	.50	—	.015
C61500	99.5	89.0–90.5	.015	—	—	—	7.7–8.3	—	1.0	—	1.8–2.2	—	—
C61800	99.5	86.9–91.0	.02	.50–1.5	—	.02	8.5–11.0	—	—	.10	—	—	—
C61900	99.5	83.6–88.5	.02	3.0–4.5	.6	.8	8.5–10.0	—	—	.10	—	—	—
C62200	99.5	83.2–86.0	.02	3.0–4.2	.6	.02	11.0–12.0	—	—	.10	—	—	—
C62300	99.5	82.2–89.5	—	2.0–4.0	.6	—	8.5–11.0	—	.50	.25	1.0	—	—
C62400	99.5	82.8–88.0	—	2.0–4.5	.20	—	10.0–11.5	—	i	.25	—	—	—
C62500	99.5	79.0–84.0	—	3.5–5.0	—	—	12.5–13.5	—	2.0	.25	—	—	—
C63000	99.5	78.0–85.0	—	2.0–4.0	.20	.30	9.0–11.0	—	1.5	.25	4.0–5.5	—	—
C63200	99.5	75.9–84.5	.02	3.0–5.0[27]	.20	—	8.5–9.5	—	3.5	.10	4.0–5.5[27]	—	—
C63400	99.5	94.9–97.1	.05	.15	.20	.50	2.6–3.2	.15	—	.25–.45	.15	—	—
C63600	99.5	93.2–96.3	.05	.15	.20	.50	3.0–4.0	.15	—	.7–1.3	.15	—	—
C63800	99.5	93.0–95.7	.05	.10	—	.8	2.5–3.1	—	.10	1.5–2.1	.10	.25–.55	—
C64200	99.5	88.2–92.2	.05	.30	.20	.50	6.3–7.6	.15	.10	1.5–2.2	.25	—	—
C64210	99.5	88.2–92.2	.05	.30	.20	.50	6.3–7.0	.15	.10	1.5–2.0	.25	—	—
C64400	99.5	88.8–91.5	.03	.05	.10	.20	3.5–4.5	—	—	.8–1.3	4.2–5.0	—	—

311

COPPER–SILICON ALLOYS (SILICON BRONZES)

Copper Alloy Number	Previous Trade Name	(18) Cu + Elements with Specific Limits, % min.	Cu (incl. Ag) % min.	Pb	Fe	Sn	Zn	Al	Mn	Si	Ni (incl. Co)
						Composition, % max. unless shown as a range or minimum					
C64700		99.5	97.0	.10	.10	—	.50	—	—	.40–.8	1.6–2.2
C64900		99.5	96.2	.05	.10	1.2–1.6	.20	.10	—	.8–1.2	.10
C65100	Low Silicon Bronze B	99.5	96.0	.05	.8	—	1.5	—	.7	.8–2.0	—
C65300		99.7	97.4	.05	.8	—	—	—	—	2.0–2.6	—
C65500	High Silicon Bronze A	99.5	94.8	.05	.8	—	1.5	—	.50–1.3	2.8–3.8	.6
C65600		99.5	94.0	.02	.50	1.5	1.5	.01	1.5	2.8–4.0	—
C65800		99.5	94.8	.05	.25	—	—	.01	.50–1.3	2.8–3.8	—
C66100		99.5	94.0	.20–.8	.25	—	1.5	—	1.5	2.8–3.5	—

MISCELLANEOUS COPPER–ZINC ALLOYS

Composition, % max. unless shown as a range or minimum

Copper Alloy Number	Previous Trade Name	Cu (incl. Ag)	Pb	Fe	Sn	Zn[20]	Ni (incl. Co)[28]	Al	Mn	Si	Other Named Elements
C66400		Rem.[20]	.015	1.3–1.7[28]	.05	11.0–12.0	.05[28]	.05	.05	.05	.02P .30–.7Co[28] .05Ag
C66700	Manganese Brass	68.5–71.5	.07	.10	—	Rem.	—	—	.8–1.5	—	—
C66800		60.0–63.0	.50	.35	.30	Rem.	.25	.25	2.0–3.5	.50–1.5	—
C66900		62.5–64.5	.05	.25	—	Rem.	—	—	11.5–12.5	—	—
C67000	Manganese Bronze B	63.0–68.0	.20	2.0–4.0	.50	Rem.	—	3.0–6.0	2.5–5.0	—	—
C67300		58.0–63.0	.40–3.0	.50	.30	Rem.	.25	.25	2.0–3.5	.50–1.5	—
C67400		57.0–60.0	.50	.35	.30	Rem.	.25	.50–2.0	2.0–3.5	.50–1.5	—
C67410		55.5–59.0	.8	1.0	.50	Rem.	2.0	1.3–2.3	1.0–2.4	.7–1.3	—
C67500	Manganese Bronze A	57.0–60.0	.20	.8–2.0	.50–1.5	Rem.	—	.25	.05–.50	—	—
C67600		57.0–60.0	.50–1.0	.40–1.3	.50–1.5	Rem.	—	—	.05–.50	—	—
C67700		55.5–58.0	.50–1.0	.7–1.5	—	Rem.	1.5–2.3	—	.05–.30	—	.40–8As
C67800		56.0–59.0	.30	.7–1.5	.20	Rem.	—	.50–1.5	.20–.6	—	—
C67810		56.5–59.5	1.0	1.0	.50	Rem.	1.5	.40–1.6	.40–1.8	.6	—
C68000	Bronze, Low Fuming (Nickel)	56.0–60.0	.05	.25–1.25	.75–1.10	Rem.	.20–.8	.01	.01–.50	.04–.15	—
C68100	Bronze, Low Fuming	56.0–60.0	.05	.25–1.25	.75–1.10	Rem.	—	.01	.01–.50	.04–.15	—
C68200		58.0–60.0	—	.6–1.0	—	Rem.	—	—	.6–1.0	.07–.15	—
C68700	Aluminum Brass, Arsenical	76.0–79.0	.07	.06	—	Rem.	—	1.8–2.5	—	—	.02–.10As
C68800		72.0–74.5	.05	.05	—	Rem.[20]	—	3.0–3.8[29]	—	—	.25–.55Co
C69000		80.0–83.0	.025	.05	—	21.3–24.1[29]	.50–.8	3.3–3.5	—	—	—
C69400	Silicon Red Brass	80.0–83.0	.30	.20	—	Rem.	—	—	—	3.5–4.5	—
C69430		80.0–83.0	.30	.20	—	Rem.	—	—	—	3.5–4.5	.03–.06As
C69440		80.0–83.0	.30	.20	—	Rem.	—	—	—	3.5–4.5	.03–.06Sb
C69450		80.0–83.0	.30	.20	—	Rem.	—	—	.40	3.5–4.5	.03–.06P
C69700		75.0–80.0	.50–1.5	.20	—	Rem.	—	—	.40	2.5–3.5	—
C69710		75.0–80.0	.50–1.5	.20	—	Rem.	—	—	.40	2.5–3.5	.03–.06As
C69720		75.0–80.0	.50–1.5	.20	—	Rem.	—	—	.40	2.5–3.5	.03–.06Sb
C69730		75.0–80.0	.50–1.5	.20	—	Rem.	—	—	.40	2.5–3.5	.03–.06P
C69800		66.0–70.0	.8	.4	—	Rem.	.50	1.4–2.3	—	.7–1.3	—
C69900	Incramute	99.5[30]	.02	.10	—	.14	.10		40.0–48.0	—	.20Co .05C .05Ag .05Cd .01As
C69910		Rem.[17]	.01	1.0–1.4	—	3.0–5.0	—	.25–.8	28.0–32.0	—	—

COPPER–NICKEL ALLOYS

Copper Alloy Number	Previous Trade Name	(18) Cu + Elements with Specific Limits, % min.	Cu (incl. Ag), % min.	Composition, % max. unless shown as a range or minimum					Other Named Elements
				Pb	Fe	Zn	Ni (incl. Co)	Mn	
C70100		99.7	—	—	.05	.25	3.0–4.0	.50	—
C70200		99.7	—	.05	.10	—	2.0–3.0	.40	—
C70400	Copper-Nickel, 5%	99.5	—	.05	1.3–1.7	1.0	4.8–6.2	.30–.8	—
C70500	Copper-Nickel, 7%	99.5	—	.05	.10	.20	5.8–7.8	.15	—
C70600	Copper-Nickel, 10%	99.5	86.5	.05[31]	1.0–1.8	1.0[31]	9.0–11.0	1.0	[31]
C70690		99.5	89.0	.001	.005	.001	9.0–11.0	.001	[32]
C70700		99.5	—	—	.05	—	9.5–10.5	.50	—
C70800	Copper-Nickel, 11%	99.5	—	.05	.10	.20	10.5–12.5	.15	—
C70900		99.5	—	.05	.6	1.0	13.5–16.5	.6	—
C71000	Copper-Nickel, 20%	99.5	—	.05	1.0	1.0	19.0–23.0	1.0	—
C71100		99.5	—	.05	.10	.20	22.0–24.0	.15	—
C71300		99.5	—	.05	.20	1.0	23.5–26.5	1.0	—
C71500	Copper-Nickel, 30%	99.5	65.0	.05[31]	.40–1.0	1.0[31]	29.0–33.0	1.0	[31]
C71580		99.5	65.0	.05	.5	.05	29.0–33.0	.30	[33]
C71590		99.5	67.0	.001	.005	.001	29.0–33.0	.001	[32]
C71700		99.5	—	—	.40–1.0	—	29.0–33.0	—	.30–.7Be
C71900		99.5	—	—	.50	—	28.0–32.0	.20–1.0	2.4–3.2Cr, .02–.25Zr, .01–.20Ti, .01C, .25Si
C72200		99.5	—	[31]	.7–1.0	[31]	15.0–18.0	1.0	[31], .30–.7Cr, .03Si, .03Ti
C72500		99.8	—	.05	.6	.50	8.5–10.5	.20	1.8–2.8Sn

COPPER–NICKEL–ZINC ALLOYS (NICKEL SILVERS)

Copper Alloy Number	Previous Trade Name	Composition, % max. unless shown as a range or minimum						Other Named Elements
		Cu	Pb	Fe	Zn[20]	Ni (incl. Co)	Mn	
C73200		70.0 min.	.05	.6	3.0–6.0	19.0–23.0	1.0	—
C73500		70.5–73.5	.10	.25	Rem.	16.5–19.5	.50	—
C73800	Nickel Silver, 70-12	68.5–71.5	.05	.25	Rem.	11.0–13.0	.50	—
C74000		69.0–73.5	.10	.25	Rem.	9.0–11.0	.50	—
C74300		63.0–66.0	.10	.25	Rem.	7.0–9.0	.50	—
C74500	Nickel Silver, 65-10	63.5–66.5	.10[34]	.25	Rem.	9.0–11.0	.50	—
C75200	Nickel Silver, 65-18	63.0–66.5	.10	.25	Rem.	16.5–19.5	.50	—
C75400	Nickel Silver, 65-15	63.5–66.5	.10	.25	Rem.	14.0–16.0	.50	—
C75700	Nickel Silver, 65-12	63.5–66.5	.05	.25	Rem.	11.0–13.0	.50	—
C75900		60.0–63.0	.10	.25	Rem.	17.0–19.0	.50	—
C76000		60.0–63.0	.10	.25	Rem.	7.0–9.0	.50	—
C76100		59.0–63.0	.10	.25	Rem.	9.0–11.0	.50	—
C76200		57.0–61.0	.10	.25	Rem.	11.0–13.5	.50	—
C76300		60.0–64.0	.50–2.0	.50	Rem.	17.0–19.0	.50	—
C76400		58.5–61.5	.05	.25	Rem.	16.5–19.5	.50	—
C76600		55.0–58.0	.10	.25	Rem.	11.0–13.5	.50	—
C76700	Nickel Silver, 56.5-15	55.0–58.0	—	—	Rem.	14.0–16.0	.50	—

COPPER–NICKEL–ZINC ALLOYS (NICKEL SILVERS) (Cont.)

Copper Alloy Number	Previous Trade Name	Composition, % max. unless shown as a range or minimum						
		Cu	Pb	Fe	Zn[20]	Ni (incl. Co)	Mn	Other Named Elements
C77000	Nickel Silver, 55-18	53.5–56.5	.10	.25	Rem.	16.5–19.5	.50	—
C77300		46.0–50.0	.05	—	Rem.	9.0–11.0	—	.01Al .25P .04–.25Si
C77400		43.0–47.0	.20	—	Rem.	9.0–11.0	—	—
C77600	Nickel Silver, 43.5-13	42.0–45.0	.25	.20	Rem.	12.0–14.0	.25	.15Sn
C78200		63.0–67.0	1.5–2.5	.35	Rem.	7.0–9.0	.50	—
C78800		63.0–67.0	1.5–2.0	.25	Rem.	9.0–11.0	.50	—
C79000		63.0–67.0	1.5–2.2	.35	Rem.	11.0–13.0	.50	—
C79200		59.0–66.5	.8–1.4	.25	Rem.	11.0–13.0	.50	—
C79300		55.0–59.0	.50–2.0	.50	Rem.	11.0–13.0	.50	—
C79600	Leaded Nickel Silver, 10%	43.5–46.5	.8–1.2	—	Rem.	9.0–11.0	1.5–2.5	—
C79800		45.5–48.5	1.5–2.5	.25	Rem.	9.0–11.0	1.5–2.5	—
C79900		47.5–50.5	1.0–1.5	.25	Rem.	6.5–8.5	.50	—

Notes

(1) These are high conductivity coppers which have in the annealed condition a minimum conductivity of 100% IACS.

(2) The value of Cu is exclusive of Ag.

(3) The total of the seven following elements—Se, Te, Bi, As, Sb, Sn, and Mn—not to exceed 40 ppm (.0040%). Hg, max., 1 ppm (.0001%); Zn, max., 1 ppm (.0001%); Cd, max., 1ppm (.0001%); S, max., 18 ppm (.0018%); Pb max., 10 ppm (.0010%); Se, max., 10 ppm (.0010%); Bi, max., 10 ppm (.0010%); oxygen, max., 10 ppm (.0010%).

(4) Includes P.

(5) Small amounts of Cd or other elements may be added by agreement to improve resistance to softening at elevated temperatures.

(6) This includes low resistance lake copper and electrolytic copper to which Ag is added.

(7) Includes B.

(8) This includes oxygen-free copper which contains P in an amount agreed upon.

(9) This includes high resistance lake copper.

(10) Te + Se.

(11) Includes Cd, deoxidized with Li or other suitable elements as a agreed upon.

(12) This includes oxygen-free tellurium-bearing copper which contains P in an amount agreed upon.

(13) Includes Te.

(14) Other deoxidizers may be used as agreed upon, in which case P need not be present.

(15) Includes Ag, S, P, and Pb.

(16) Includes oxygen-free or deoxidized grades with deoxidizers (such as P, B, Li, or other) in an amount agreed upon.

(17) Total named elements shall be 99.8% min.

(18) Specific limits are defined as any numerical values, whether maximum only, minimum only, or ranges.

(19) Ni + Co, .20% min.; Ni + Fe + Co, .6% max.

(20) These specification limits do not preclude the possible presence of other unnamed elements. However, analysis shall regularly be made for the minor elements listed in the table plus all major elements except one. The major element which is not analyzed shall be determined by difference between the sum of those elements analyzed and 100%. By agreement between producer and consumer, analysis may be required and limits established for elements not specified.

(21) Al + Si, .005% max.

(22) For tube over 5 in. O.D., Pb may be less than .20%.

(23) Copper, 61.0% min. for rod.

(24) Copper, 60.0% min. for rod.

(25) For flat products, the minimum Sn content may be .8%.

(26) Includes Zn.

(27) Fe content shall not exceed Ni content.

(28) Fe + Co shall be 1.8-2.3%. Co not included with Ni.

(29) Al + Zn shall be 25.1-27.1%.

(30) Total named elements shall be 99.5% minimum.

(31) When the product is for subsequent welding applications and so specified by the purchaser, Zn shall be .50% max., Pb .02% max., P .02% max., S .02% max., and C .05% max.

(32) The following maximum limits shall apply: .03C, .02Si, .003S, .002Al, .001P, .0005Hg, .001Ti, .001Sb, .001As, .001Bi, and .001Sn.

(33) The following maximum limits shall apply: .07C, .15Si, .024S, .05Al, and .03P.

(34) Pb, .05% max. for rod and wire.

INDEX

Acetic acid, 109
Air-conditioned pulpits, 15
Alloys list, 46, 303-316
Aluminum, 44, 45, 58
  cuprous oxide film and, 108, 111
  prefix designation, 44
Aluminum bronzes, number designations,
  45
American National Standards Institute
  (ANSI), 264
American Society of Mechanical Engineers
  (ASME), 264
American Society for Testing and
  Materials (ASTM), 43, 48, 70-81
  specifications, 263-278
Anisotropy, 288
Annealing, 25-34, 279-280
  basic purpose of, 33
  batch, 28, 155
  coil, 28-31
  conduction and, 28
  continuous strand, 31-33
  strip, 31-33
  temper and, 47-49
  temperature and, 25-34
Anodic cleaning, 144
Aqueous corrosion, 109-112, 114-116
Arc welding, 191, 192, 193, 194, 195
ASM Metals Handbook, 147, 148
Atmospheric oxidation, 31, 32
  corrosion and, 112-114
  field experience, 112-114
  protection from, 4
Atomic structure, 51-53
Atomic vibration, 89
Atoms:
  dislocations, 55
  substitutional solid solutions, 56

Batch annealing, 28, 155
Bedworth, R. E., 108, 131
Bell furnaces, 28-31
Bend curves, 175, 176, 184, 186, 187
Bend orientation, 242
Bend radius, 174-177, 184-187
Bending:
  forming operations, 174-180, 240-243
  reversed, 235
  springs, 240-243
Benzotriazole, 34
Beryllium coppers, 60-61
Beta-ray instruments, 22
Binary alloy systems, 62
Bio-fouling, 118-119
Blanking, 41, 163
Blisters, 282
Body-centered cubic structures, 56-57
Bohr-Sommerfeld theory, 51
Bonding:
  eutectic, 90
  vs. soldering, 198
Bonds, metallic, 52
Book molds, 5
Brasses:
  cartridge, 24, 44
  color, surface treatment, 212
  coloring by chemical treatment,
    213-219
  density, 92
  high zinc, 24, 146, 154-155
  lacquered, 213
  leaded, 44-45, 59
  number designations, 44-45
  scrap, 4, 63
  specifications, 80-81
  tin, 45
    specifications, 80-81

Brazing, 191-196
  operating data for, 197
Breakaway velocity, 119
Bright dip solutions, 108, 109
Brinell hardness test, 72
Bronzes:
  aluminum, number designations, 45
  color, surface treatment, 212
  lacquered, 213
  leaded, 10
  phosphor, 10, 24, 34, 146
    coefficients of friction, 155
    leaded, 45
    number designations, 45
    slitting, 37
    specifications, 80-81
  silicon, number designations, 45
Brushes, 34
Buffability, 280
Buffing, 219, 221-223
Bulging, 159
Burrs, 39

Cadmium plating, 140
Camber, 35
  tolerance specifications, 273, 276-277
Cantilever beams, 247, 252, 256, 259, 273,
  276
Cartridge brass, 24
  number designation, 44
Casting, 3-12
  direct chill, 7-10, 11
  horizontal continuous, 8, 10, 11
  two basic operations, 3
Cathodic cleaning, 140-141
Charge buckets, 4-6, 8
Chloride type fluxes, 200-201
Chromium plating, 140
Circumferential tension, 181
Classification systems, 43-50
  Copper Development Association, Inc.
    (CDA), numbering system, 43-46
  prefix system, 44
  temper designations, 46-50, 79-81
Cleaning, 34, 109
  anodic, 144
  cathodic, 140
Cleaning equipment, 108
Cluster rolling mills, 18-21
Cobalt, 46
Coil annealing, 28-31
Coil set, 38-39

Coining, 143, 145, 148, 160-161
  springs, 240
Cold rolling to final thickness, 17-25
Color, surface treatments, 211-227
Coloring by chemical treatment, 213-218
Comparison procedure, 71-72
Compression stress, 231
Compressive yield strength, 231
Computer use:
  process-control, 28
  springs, 246
Conduction:
  annealing and, 28
  impurities and, 62-63
Conductivity, see Electrical conductivity;
    Thermal conductivity
Constitution diagrams, 57
Contact force, springs, 231-235, 247-251
Contact resistance, see Electrical
    contact resistance
Continuous strand annealing, 31-33
Copper Development Association (CDA),
    43, 213, 214
  numbering system, 43-46
Copper oxide, 7
Copper zinc, 56
  in coining, 145
  color, surface treatment, 231-235
  number designations, 44-45
Corrosion, 107-112
  aqueous, 109-112, 114-116
  atmospheric oxidation and, 112-114
  fretting, 281
  galvanic, 116-118
  intergranular, 121, 129, 131
  springs, 236-237
  stress, 122-128
  transgranular, 123
Corrosion cells, 110
Corrosion fatigue, 128-129
Corrosion rates:
  in brine, 117
  in cascading sea water, 118
  conversions, 116
  instantaneous, 117-118
  submerged in sea water, 118
Corrosion resistance, 1, 51, 55, 89, 107-131,
    146
  bio-fouling, 118-119
  corrosion fatigue, 128-129
  dezincification, 129, 130
  erosion-corrosion, 119-121

inlet, 122
  velocity and, 119-121
field experience, 112-116
  aqueous corrosion, 109-112, 114-116
  atmospheric oxidation, 107-108,
    112-114
  galvanic, 116-118
  intergranular, 124, 129, 131
  pitting, 120-122
  stress, 122-128
Cost factors, 285, 302
  basic elements, 285, 286
  comparative, methods for calculating,
    301-302
  cost-effectiveness, 262
  density, 286-287
  directionality, 288-296
  equation for, 302
  length, 287-288
  springs, 243-244
  strength, 297-301
Cracking:
  gas units, 29
  season, 122
Creep, 96-100
  strength, 100
  temperature and, 101, 102
Creep curve, 100, 102
Cryogenic temperatures, 88
Crystallization, by nucleation, 52, 53
Crystals:
  aggregations of, 70
  imperfections, 55
  orientations, 70-71
  structure of, 55-59
    body-centered cubic, 54-57
Cupro nickels:
  coefficients of friction, 154
  number designations, 46
Cuprous oxide, 107-108, 111
Cuprous oxide film, 107-108, 111
Cutting, 34, 41

Deep drawing, 147-151, 158
Deflection:
  electrons, 83, 85
  springs, 229-231, 247-256
  stress and, 251-256
Deformation:
  elastic, 180-181
  plastic, 53-55, 180-181
Degreasing, 108, 155

Densities, 91-92
  as cost factor, 286-287
Detergents, 34
Dezincification, 129, 130, 279
Diamond cone penetrator, 77
Die block, 156-157
Dip coating, 136-139
Direct chill casting, 7-10, 11
Direct chill slabs, hot rolling, 12-17
  furnaces, 12-17
  television in, 15-17
  temperature in, 12-17
Directionality:
  as cost factor, 288-296
  grain size and, 28, 71
Drag, 157
Drawability, 148
Drawing operations, 158-159
  bulging, 159
  deep drawing, 147-151, 158
  defects in, 171-173
  grain size and, 147-151
  ironing, 158-159
  lubrication, 167-171
  material design considerations, 165-166
  necking, 160
  in presses, 156-158
  redrawing, 158, 166
  reverse redrawing, 158
  shallow drawing, 158
  sizing, 159
Drawing tools, 156-157
Drifting, 164-165
Dry sliding friction, 235
Ductile-brittle transition, 237
Ductility, 24-25, 51, 53, 145
  reduction of, 55
Dynamic contact resistance, 133

Edge dislocation, 55
Edge rolling, 42
Elastic deformation, 180-181
Elastic limit, 67, 68
Elastic modulus, 67-69
Electric induction melting furnaces, 4, 6
Electric conductivity, 1, 51, 53, 60, 82-88
  conversion formulas, 89
  impurities and, 85-88
Electrical contact resistance, 132-142
  dynamic, 133
  plating, 135-142
    materials for, 135, 136

methods of, 136-138
static, 132-133
Electrochemical mechanisms, 109
Electroless chemical plating, 142
Electron flow, 53, 109-111
Electron shells, 51, 83
Electrons, 83, 85
valence, 52
Electroplating, 139
Electroplating Engineering Handbook, 142
Elongation, 69, 149
grain size and, 147-149
Embossing, 148, 160-161, 225-227
Endurance limit, 95
Engineering strain, 66
Engineering stress, 66
Equiaxed grain size, 71
Erosion-corrosion, 119-120
inlet, 122
velocity and, 119-121
Eutectic bonding, 90

Fabrication, 143-189
drawing operations, 158-159
bulging, 159
deep drawing, 147-151, 158
defects in, 171-173
grain size and, 147-151
ironing, 158-159
lubrication, 167-171
material design considerations,
165-166
necking, 160
in presses, 156-158
redrawing, 158, 166
reverse redrawing, 158
shallow drawing, 158
sizing, 159
forming operations:
bending, 178-180, 240-243
blanking, 41, 163
coining, 143, 145, 148, 160-161, 240
drifting, 164-165
hole flanging, 164-165
spinning, 164
springs, 240-243
stretch draw forming, 161, 162
stretch forming, 161-162
Face-centered cubic lattices, 53
Fatigue:
corrosion, 128-129
testing, 94-96

Fatigue strength, 235-236
Ferric sulfate, 109
Flanging, hole, 164-165
Flatness tolerance specifications, 277
Flattening rolls, 40
Flexural stress, 231
Fluorescence, x-ray, 9
Fluxes, 197, 204-205, 209
comparison of, 196-197
Formability, 1
grain size and, 71
Forming operations:
bending, 178-180, 240-243
blanking, 41, 163
coining, 143, 145, 148, 160-161, 240
drifting, 164-165
hole flanging, 164-165
spinning, 164
springs, 240-243
stretch draw forming, 161, 162
stretch forming, 161, 162
Four-high rolling mills, 20-22
Fretting corrosion, 281
Friction, 162
coefficients of, 155
dry sliding, 235
Furnaces:
bell, 28-31
electric induction melting, 4, 6
holding, 5, 6
in hot rolling of direct chill slabs, 12-14
Fusion welding, 191, 192, 193, 194

Galvanic corrosion, 116-118
Galvanic series in sea water, 119
Gas cracking units, 29
Gauge, 17-23
specifications, 278
Gauge length, 69
Gilding metal, 212
Gold, 51
plating, 135-136, 137-141, 212
Grain boundaries, 70-72
Grain size, 26-28, 70-74
control of, 26, 28, 72, 74
directionality and, 28, 72
drawing operations and, 147-151
elongation and, 148-149
equiaxed, 71
formability and, 70
measurement of, 71-74
ranges and recommended applications, 153

specifications, 147
strain and, 26
strength and, 26, 70
surface and, 25, 71, 151-153
temper and, 47-48
tensile strength and, 144, 145
thickness and, 149
workability and, 28
Gravity, specific, 91-92
Grip strength, 234

Hardness, 24, 55
strength and, 47
testing, 72, 75-78
Heyn intercept method, 71-72
High-speed coil milling machine, 18
Holding furnaces, 5, 6
Hole flanging, 164-165
Hooke's law, 67
Horizontal continuous casting, 8, 10, 12
Hot rolling of direct chill slabs, 12-17
furnaces, 13, 14
television in, 15-17
temperature in, 14
Hydrochloric acid, 108, 109, 215, 216, 218, 236

Impurities, 62-64
conduction and, 62-63
electrical conductivity and, 85-88
solubility of, 85
Inlet erosion-corrosion, 120
Instantaneous corrosion rates, 117-118
Inter-crystalline corrosion, 129
Intergranular corrosion, 121, 129, 131
Inter-metallic compounds, 59
International Annealed Copper Standard (IACS), 83
International Standards Organization (ISO), 44, 264
International System of Units (SI), 66
Iron:
cuprous oxide film and, 108, 111
as an impurity, 64, 86
Ironing, 158-159

Jeffries planimetric procedure, 71-72
Joining, 190-210
defined, 190
selection guide for processes of, 191-194
see also Brazing; Soldering; Welding

Knife sets, 34-38, 270
Knoop micro-hardness test, 72, 75
Knurled finish, 226-227

Lacquers, 219
Laminations, 282
Latin America, metal designations, 44
Lattices, face-centered cubic, 53
Lead, 56, 59
plating, 140
Leaded brass, 59
number designations, 44
Leaded bronze, 10
Leaded nickel silver, 46
Leaded phosphor bronze, 45
Length:
as cost factor, 288-290
tolerance specifications, 272-273, 274, 275
Leveling, 34, 36-40, 277
Looping towers, 31
Lubricants, 22-23
oil-base, 168-169
over-all economy, 170
selection guide, 167-171
stability in storage, 170
universal factors, 170-171
water-base, 167-168
Lubrication, 34, 155, 278-281
drawing operations, 167-171

Machinability, 59
Manganese, 46
Manufacturing Processes and Materials for Engineers, Doyle/Keyser/Leach/ Schrader/Singer, 189
Mattson's solution, 125
Mean free path, 85
Mechanical properties, 65-81
Meganewtons, 66
Megapascals, 66
Melting, 3-12
Metal dipping, 136-139
Metal insert-gas (MIG) welding, 191
Metallic bonds, 52
Metallurgy, 51-64
and alloying effects, 55-57
atomic structure, 51-53
crystal structure, 51-53
impurities, effect of, 62-64
phase diagrams, 57-62
plastic deformation, 53-55, 180-181
precipitation hardening, 57-62
slip, 53 55, 56

Mill processing operations, 3-42
  annealing, 25-34, 279-280
    basic purpose of, 33
    batch, 28, 155
    coil, 28-31
    conduction and, 28
    continuous strand, 31-33
    strip, 31-33
    temper and, 47-49
    temperature and, 25-34
  casting, 3-12
    direct chill, 7-10, 11
    horizontal continuous, 7, 10, 12
    two basic operations, 3
  cleaning, 36, 109
  cleaning equipment, 108
  cold rolling to final thickness, 17-25
  cutting, 34, 41
  hot rolling of direct chill slabs, 12-17
    furnaces, 13, 14
    television in, 15-17
    temperature in, 14
  leveling, 34, 36-40, 277
  melting, 3-12
  milling, 17
  raw materials, 3-5
  scalping, 17
  slitting, 34-38
Milling, 17
Molds, 5-12
  book, 5
  sizes of, 5, 6, 7, 9, 10

Necking, 69, 160
Nickel, 44, 58, 89-91
  cubic structure of, 58
  cupro:
    coefficients of friction, 155
    number designations, 46
  cuprous oxide film and, 111
  diffusion barrier, 135
  as an impurity, 87
  plating, 136, 137, 138, 140, 142
  prefix designation, 44
Nickel irons, 89-90
Nickel silvers, 24
  coefficients of friction, 155
  color, surface treatment, 212, 213
  leaded, 46
  number designations, 46
Nicotera, E. T., 142
Nitric acid, 108, 109, 215, 218

Nobility, 1, 51
Nucleation, crystallization by, 52, 53
Numbering system, 43-46

Offset yield strength, 69, 187, 230
Oil-base lubricants, 168-169
Olin Corporation, 112, 132, 256, 257, 258, 260, 298
Organic type fluxes, 197
Over-arm separators, 36
Oxidation, atmospheric, 31, 32
  corrosion and, 112-114
  field experience, 112-114
  protection from, 4
Oxide scumming, 5
Oxyacetylene welding, 190

Palladium plating, 136
Patent leveling, 277
Periodic table of elements, 51
Phase of an alloy system, 56
Phase diagrams, 57-62
Phosphor bronzes, 10, 24, 34, 146
  coefficients of friction, 155
  leaded, 45
  number designations, 45
  slitting, 37
  specifications, 80-81
Phosphoric acid, 109
Phosphorus, 45
Physical properties, 65, 82-93
Pickling solution, 109
Pilling, N. B., 131
Pitting, 120-122
Plane strain, 179, 180, 181
Plastic deformation, 53-55, 180-181
Plastic flow, 70
Plating, 133-142, 203-210
  materials for, 135-136
  methods of, 136-142
Polishing, 220-225
Pouring boxes, 5-7, 9
Precipitation hardening, 57-62
"Predicting Contact Force, Deflection, and Stress in a Terminal Connector Design", Zarlingo, 247, 260
Prefix system, 44
Pressure pad, 156
Process-control computers, 28
Profilometers, 220
Proportional limit, 67, 68, 69
P-type semiconductors, 111

Pulpits, air-conditioned, 16
Punches, 156, 159, 160-163

Quality, *see* Specifications
Quaternary system, 62

Raw materials, 3-5
Recrystallization, 17, 33
  temper and, 47
  temperature and, 12, 25, 33
Redrawing, 158, 166
Reflectance:
  electrons, 83, 85
  spectral, 51
Resin type fluxes, 197
Resistance welding, 191, 192, 193, 194
Resistivity:
  conversion formulas, 89
  temperature and, 88
  see also Electrical contact resistance
Reverse bending, 235
Reverse redrawing, 158
Rockwell hardness test, 72-78
Rockwell superficial hardness test, 77, 78
Roll grinding shops, 23
Rolled patterns on strip, 225-227
Rolling mills, 17-25
  cluster, 19-21
  four-high, 19, 20
Rosin type fluxes, 197
Rotary-knife-slitters, 270-271

Scalping, 17
Scrap, 3-5
  baled, 4
  brass, 4, 63
  web, 296
Season cracking, 122
Secant modulus, 68, 231-232, 248-249,
  268-269
Semiconductors, 90, 111
Sendzimir mills, 19-21
Shallow drawing, 158
Shear stress, maximum, 54
Shelf-life solderability, 202-203
Shell electrons, 51, 53
Shielded metal-arc welding, 191
Silicon, 90
  as an impurity, 85, 87
Silicon bronzes, number designations, 45
Silicon carbide plates, 10

Silver, 51
  plating, 139, 140
Single-phase solid-solution alloys, 56
Sizing, 159
Slip, 53-55, 56
Slip planes, 53-54
Slitting, 34-38
Slivers, 282
S-N curves, 95, 235-236
Society of Automotive Engineers (SAE), 43
Sodium cyanide, 109
Sodium dichromate, 109
Solderability, 148, 201, 203, 206, 212
  ratings, 203, 204-209
  relative, 203
  shelf-life, 202-203
  testing, 201
Soldering, 109
  methods of, 201-203
  vs. other bonding methods, 196
Solders, 196, 199
Solid-solution alloys, 56
Solid-solution strengthening effect, 56
Solid-state welding, 192, 193, 194
Solution velocity, corrosion rates in brine
  and, 117
Specific gravity, 91-92
Specifications, 79-80, 261-284
  alloy (composition), 278
  camber tolerance, 276-277
  flatness tolerance, 277
  gauge, 278
  grain size, 147
  length tolerance, 272-273, 274, 275
  organizations publishing, 263-264
  quality, description of, 261-262
  straightness tolerance, 273, 276-277
  surface, 278-283
  temper, 278
  thickness tolerance, 265-270
  trouble shooting, 283-284
  width tolerance, 270-272
Spectral reflectance, 151
Spectrographic laboratory, 8
Spinning operations, 164
Spring Designer's Data Package, The, 260
Springback behavior, 188
Springs, 228-260
  application factors, 245-256
    computer use, 246
    contact force, 247-248
    deflection, 247-254

restrictions, 256
selection model, 245, 246
stress, 251-254
stress relaxation, 254-256
cost factors, 243-244
design criteria, 228-244
forming operations, 240-243
bending, 240-243
coining, 240
formulas, 258-259
mechanical criteria, 229-236
contact force, 231-235, 247-251
corrosion, 236-237
deflection, 229-231
life cycles, 235-236
stress relaxation, 254-256
temperature effects, 237-239
new data, 260
testing, 256-258
Squeegee rolls, 36
Static contact resistance, 132-133
Straightness tolerance, specifications, 273, 276-277
Strain:
engineering, 66
grain size and, 26
measurement of, 66
plane, 179, 180, 181
-stress diagram, 67, 68
Strength, 1, 23-25
as cost factor, 297-301
creep, 100
fatigue, 235-236
grain size and, 26, 70
grip, 234
hardness and, 47
increasing, 55-57
solid-solution effect, 57
tensile, 24, 49, 66-70, 79-81
grain size and, 144, 145
ultimate, 183
yield, 66-69, 187, 236
compressive, 231
offset, 230
tensile, 230
Stress, 1
compression, 231
deflection and, 251-256
engineering, 66
flexural, 231
locked-in, residual, 123-124
measurement of, 66

relieving, 34, 35, 125
shear, maximum, 54
springs, 230-231, 247-254
tensile, 230
Stress corrosion, 122-128
Stress relaxation, 103-105
springs, 233-234, 254-256, 274-276
Stress rupture, 96, 100-103
Stress-strain diagram. 66-69
Stretch draw forming, 161, 162
Stretch forming, 161, 162
Stretcher leveling, 277
Strip annealing, 31-33
Substitutional solid solutions, 56
Sulfuric acid, 109, 215, 217, 218, 235, 236
Surface, 153-156
grain size and, 25, 71, 151-153
preparation of, 200
specifications, 278
stain, 281-282
treatments, 211-227
buffing, 219, 221-223
color, 211-227
coloring by chemicals, 213-218
polishing, 219-223
rolled patterns on strip, 225-227

Tandem mills, 19, 20
Tarnish inhibitors, 34
Television in hot rolling of direct chill slabs, 15-17
Temper, 46
annealing and, 47-49
common names, 49
designations, 46-50, 79-80, 82-84
grain size and, 47-48
recrystallization and, 47
specifications, 278
Temperature:
annealing and, 25-34
creep and, 101, 102
cryogenic, 88
in hot rolling of direct chill slabs, 14
recrystallization and, 12, 25, 33
resistivity and, 88
springs, effect of, 237-239
thermal expansion and, 90-91
Tensile strength, 23-24, 49, 69-70, 79-80
grain size and, 144, 145
ultimate, 183
Tensile stress, 230
Tensile yield strength, 230

Tension:
  circumferential, 179, 181
  testing, 66-70
  transverse, 179, 181
  uniaxial, 180, 181
Ternary system, 62
Texture, 181-184
Thermal conductivity, 1, 51, 53, 57, 89-90
Thermal expansion, 90-91
  temperature and, 90-91
Thickness:
  grain size and, 149
  tolerance specifications, 265-270
Tin, 56
  color, surface treatment, 212
  cuprous oxide film and, 108, 111
  plating, 136-139
Tin brasses:
  number designations, 45
  specifications, 80-81
Tin lead, plating, 137, 138, 139
Tolutriazole, 34
Transgranular corrosion, 123
Transverse tension, 179, 181
Trisodium phosphate, 218
Tungsten inert-gas (TIG) welding, 195

Uniaxial tension, 180, 181
Unified Numbering System (UNS), 44-46
Unit cell construction, 52
United States penny, 145

Valence electrons, 52
Velocity:
  breakaway, 119
  erosion-corrosion and, 119-120
  solution, corrosion rates in brine and,
    117
Vibration, atomic, 89

Vickers hardness test, 72

Water-base lubricants, 167-168
Wear resistance, 55
Web scrap, 296
Welding, 190-191
  arc, 191-195
  fusion, 191, 192, 193, 194
  metal inert-gas (MIG), 191
  oxyacetylene, 190
  resistance, 191-194, 198
  shielded metal-arc, 195
  solid-state, 192, 193, 194
  tungsten inert-gas (TIG), 195
Width tolerance specifications, 270-272
Wiedeman and Franz, law of, 89-90
Work-hardening rates, 146
Wrought products, 12

X-ray fluorescence, 9

Yield locus, 180-181, 183
Yield strength, 66-69, 187, 236
  compressive, 230
    flexural, 230
    offset, 230
    tensile, 230
Young's modulus, 67-69, 232, 247-254, 256

Z mills, 21-23
Zarlingo, S. P., 204-205, 218-220, 247
Zinc, 33, 56
  cubic structure of, 58-59
  cuprous oxide film and, 108, 112
  density, 91-92
  plating, 140
  weight percentage, 59
Zinc brasses, 24, 146
  coefficients of friction, 154-155
Zinc oxide, 33, 280